"Layer after layer, this edited book unfolds outbound Chinese tourism experiences in various countries. By doing so, it provides informed insights on how to attract, accommodate, and serve this burgeoning market."

—*Jafar Jafari, University of Wisconsin-Stout, USA*

"This book maps the evolution of Chinese outbound travel. It is an excellent resource–for individuals interested in tourism and hospitality, including domain experts, researchers, and students. Industry stakeholders and general readers alike will appreciate this easy-to-digest overview of a globally promising tourist segment."

—*Dingbin Long, Consul General of The People's Republic of China in Perth, Australia*

"Given the dynamic and ever-expanding nature of the Chinese market, there is heightened interest in further understanding the structure and changing characteristics of such an important global force. This is a timely and comprehensively crafted book to touch not only on the inner layers of outbound travel behavior, but also shed light on the emerging trends of Chinese tourists. It is a must have book for researchers and practitioners."

—*Muzaffer Uysal, University of Massachusetts–Amherst, USA*

"This book is a must read for anyone with an interest in the outbound tourist market from China. Much of what is reported about this market is purely numerical; so, this text provides the much-needed greater depth about Chinese tourist behavior abroad. Scholars and practitioners will find the contents on market segmentation to be particularly enlightening and useful."

—*Alastair M. Morrison, University of Greenwich, UK*

"This book provides a holistic picture of the Chinese outbound tourism in the past, present, and future. It is a highly desirable resource for tourism and hospitality scholars, students, and industry practitioners worldwide to better understand Chinese outbound tourism and tourists' behavior and experiences."

—*Fang Meng, University of South Carolina, USA*

"This book provides a superb overview of the emerging Chinese outbound tourism industry. The editors' intimate understanding of the market is showcased through their exposition of the history of the Chinese tourism market, followed by clear and well thought out insights into emerging trends and factors which influence consumer behavior within the tourism market. On all accounts, I highly recommend this book."

—*Ian Phau, Curtin University, Australia*

"By introducing cutting edge research, this book offers a meticulous overview of the Chinese outbound tourism market development and tourist behaviors. The editors provide a pragmatic and analytical look at this large tourism source country in the world, which is a useful reference book for practitioners and researchers."

—*Danni Zheng, Fudan University, China*

Chinese Outbound Tourist Behaviour

Assuming an international perspective, *Chinese Tourist Outbound Behaviour* presents an insightful exploration of the evolution of China's tourism market, explores Chinese tourists' behaviour, and considers how the country's tourism landscape will expand in the future.

Featuring 17 chapters compiled and written by industry experts representing 11 countries, this collection offers a vivid profile of Chinese tourists and the characteristics distinguishing them from other market segments. This book coincides with the growing interest in Chinese tourism and tourist behaviour as the top market in the world in terms of tourism spending and arrival numbers, presenting an overview of Chinese tourist segments and travel-related concerns to paint a clear picture of the market's status. Chapters address the future of Chinese tourism, providing industry stakeholders an up-to-date view on this valuable market along with suggestions to best harness the market's power.

Providing an up-to-date exploration of numerous contemporary issues, this book will be valuable to a wide audience, including advanced students in tourism, hospitality and leisure and recreation studies and stakeholders, authorities, establishments and employees within the tourism industry. This book offers readers greater knowledge about the past, present and future of the Chinese outbound tourism market.

Jun Wen is a lecturer in Tourism and Service Marketing at the School of Business and Law, Edith Cowan University (ECU), Australia. Jun is an award-winning early career researcher. He was named among the Top 40 of Australia's early achievers (Rising Stars) in 2020 and 2021.

Metin Kozak holds Ph.D. degree from Sheffield Hallam University, UK. He is elected as the Fellow of the *International Academy for the Study of Tourism*, and *Tourist Research Centre.* His research interests entail marketing and consumer behavior. His current affiliation is with Kadir Has University, Turkey and he is the co-editor of the journal *Anatolia*.

Routledge Advances in Tourism
Edited by ***Stephen Page****, Hertfordshire Business School, University of Hertfordshire, UK*

Hosting the Olympic Games
Uncertainty, Debates, and Controversy
Edited by Marie Delaplace and Pierre-Olaf Schut

Encountering Nazi Tourism Sites
Derek Dalton

Evaluating the Local Impacts of the Rio Olympics
Edited by Marcelo Neri

Mountain Resort Marketing and Management
Armelle Solelhac

Ageing and the Visitor Economy
Global Challenges and Opportunities
Stephen J. Page and Joanne Connell

Asian Mobilities Consumption in a Changing Arctic
Edited by Young-Sook Lee

Tourism and Foreign Direct Investment
Issues, Challenges and Prospects
Edited by H. Cristina Jönsson

Chinese Outbound Tourist Behaviour
An International Perspective
Edited by Jun Wen and Metin Kozak

For more information about this series, please visit: www.routledge.com/advances-in-tourism/book-series/SE0538

Chinese Outbound Tourist Behaviour

An International Perspective

Edited by
Jun Wen and Metin Kozak

Routledge
Taylor & Francis Group
LONDON AND NEW YORK

First published 2022
by Routledge
2 Park Square, Milton Park, Abingdon, Oxon OX14 4RN

and by Routledge
605 Third Avenue, New York, NY 10158

Routledge is an imprint of the Taylor & Francis Group, an informa business

British Library Cataloguing-in-Publication Data
A catalogue record for this book is available from the British Library

Library of Congress Cataloging-in-Publication Data
A catalog record has been requested for this book

ISBN: 978-0-367-63918-1 (hbk)
ISBN: 978-0-367-63920-4 (pbk)
ISBN: 978-1-003-12132-9 (ebk)

DOI: 10.4324/9781003121329

Typeset in Times New Roman
by codeMantra

Contents

Figures

Tables

Contributors

Vedat Acar is an Assistant Professor in the School of Tourism, Aydin Adnan Menderes University, Turkey. He has been a professional licenced tour guide since 2009. He is specialized in tourist behaviour and tour guiding. He has been conducting qualitative research by using such data collection tools as participant observation, focus group and semi-structured interviews since five years.

Wolfgang Georg Arlt FRGS FRAS has been working for all his academic and professional life with China and with tourism. He is the founder and CEO of COTRI China Outbound Tourism Research Institute and continues to work as consultant, trainer and researcher based in Hamburg/ Germany.

Vorakarn Chalermchaikit is a Ph.D. candidate majoring in Hospitality and Tourism Management at Assumption University, Thailand. Currently, she is a part-time lecturer majoring in service industry. Her research interests focus on gender issue in service industry, cross-cultural differences in hospitality and tourism and sustainability and tourist behaviour.

Yixuan Cheng is a Ph.D. candidate, majoring in Hospitality and Tourism Management at Assumption University, Thailand. Her research interests include rural tourism, community tourism and outbound tourism.

Yedan Fan is a Ph.D. candidate in the Graduate School of Business at Universiti Sains Malaysia, Malaysia. Her research area is about Chinese guest behaviour, hospitality marketing, consumer emotion and consumer experience.

Elspeth Frew is an Associate Professor in Tourism, Hospitality and Event Management in the La Trobe Business School, Latrobe University, Melbourne. Elspeth's research interest is in various aspects of tourism management, with a particular focus on heritage management and attraction and event management.

Shuaibu Chiroma Hassan is a graduate of the University of Bedfordshire, UK; Usmanu Dan Fodiyo University, Sokoto, Nigeria and Kaduna Polytechnic, Nigeria. He is currently a Senior Lecturer in Hospitality and

Tourism Management and Chief Coordinator of Students' Work Experience Scheme (SIWES).

Fangli Hu received the M.S. degree from Tourism Management, Fudan University, Shanghai, China (2021). Her research interests are tourist behaviour, tourism destination marketing, big data in tourism and quantitative research methods.

Lala Hu is an Assistant Professor of marketing at the Catholic University of Milan, Italy. Her research interests include international marketing, cross-cultural marketing and tourism marketing, with a special focus on the Chinese market.

Foad Irani is a Ph.D. candidate and research assistant in the Faculty of Tourism Eastern Mediterranean University, North Cyprus, Turkey. His research areas include green tourism, sustainable tourism and responsible travel, consumer behaviour and human resource management.

Salmi Mohd Isa is an Associate Professor and Deputy Dean of Academic, Career and International at Graduate School of Business, Universiti Sains Malaysia, Malaysia. Her research area focuses on corporate social responsibility, tourism marketing, neuromarketing, marketing innovation and branding.

Matias Thuen Jørgensen is an Associate Professor and head of the Centre for Tourism Research at Roskilde University, Denmark. He holds a Ph.D. in tourism management from the School of Hotel and Tourism Management, The Hong Kong Polytechnic University, SAR China. Matias has done extensive work on China Outbound Tourism.

Anders Ellemann Kristensen holds an M.A. in Chinese and a Graduate Certificate in Business Administration from Aarhus University, Denmark. He is the General Manager at Albatros Travel China, is part of the EU working group for tourism and is sitting on the board of the EUSME Center in Beijing.

Clare Lade is a Lecturer in the Bachelor of Hospitality Management at Melbourne Polytechnic, Australia. Her research interests include regional tourism development, international tourism, gastronomic tourism and festival and event management.

Alastair M. Morrison is currently a Research Professor in Marketing, Events and Tourism at the Business School, University of Greenwich, UK. He is the Co-Editor-in-Chief of the *International Journal of Tourism Cities*.

Mirko Olivieri is a Ph.D. candidate in Communication, Markets and Society at the Department of Business, Law, Economics and Consumer Behavior of IULM University, Milan, Italy. His research interests lie in the fields of marketing communication and corporate communication, with a focus on digital channels management.

Jianrong Peng is a researcher at the Center for Latin American and Caribbean Studies, Southwest University of Science and Technology, Sichuan, China. She is also a Ph.D. candidate at the Tourism Research Institute, University of Alicante, Spain. Her main line of research focuses on Chinese outbound tourism, with a special emphasis on Latin America as a destination.

Diego Quer is an Associate Professor at the Department of Management, University of Alicante, Spain, and a visiting professor at the Center for Latin American and Caribbean Studies, Southwest University of Science and Technology, Sichuan, China. His research interests focus on international management and tourism in the Chinese context.

Sima Rahimizhian is a Ph.D. candidate and research and teaching assistant in the School of Computing & Technology Eastern Mediterranean University, North Cyprus, Turkey. Her research interest includes information technology management, technology in travel and tourism, innovations in hospitality and tourism management, internet and social media marketing.

Ibrahim Sani is a graduate of Tourism from the Institute of International Tourism, Varna, Bulgaria. He has an M.B.A. from Usmanu Dan Fodiyo University, Sokoto, Nigeria. He is Director of Tourism, Ministry of Commerce and Industries, Katsina State, Nigeria.

Noel Scott is an Adjunct Professor of Tourism Management in the Sustainability Research Centre, University of the Sunshine Coast, Australia. His recent focus is on the use of cognitive psychology theory and psychophysiological methods to understand tourist experiences. He is a Fellow of the Council of Australian University Tourism and Hospitality Educators and a member of the International Association of China Tourism Scholars.

Nitasha Sharma is a tourism geographer working at the Faculty of Spatial Sciences, University of Groningen, the Netherlands. Her research broadly examines the multiple and contested representations of South Asia and Europe through projects situated in critical tourism studies and human-environment interactions. Her research interests are dark tourism, sustainable tourism, authenticity, moral issues in tourism, tourist psychology and behaviour, rituals and sacred spaces and the role of power relations in tourism encounters within different political and socio-cultural contexts.

Han Shen is a Professor of Tourism Management at Fudan University, China. Her research interests are tourism consumer behaviour, tourism destination marketing and management, new media marketing and advertising.

Tony S. M. Tse is a Professor of Practice in the School of Hotel and Tourism Management at The Hong Kong Polytechnic University, Hong Kong SAR, China. He has research expertise in China outbound tourism, tourism marketing and hospitality management education.

Sera Vada is a Research Fellow with the International WaterCentre, Griffith University, Australia. Her research and consulting interests include positive psychology and tourist behaviour studies, destination marketing and diversification of tourist markets, tourism challenges in the Pacific and Chinese tourism.

Sina Vahedi is a Research Assistant in the Faculty of Tourism Eastern Mediterranean University, North Cyprus, Turkey. His research areas include consumer behaviour, human resource management, accounting and marketing management.

Jun Wen is a Lecturer in Tourism and Hospitality Management in the School of Business and Law at Edith Cowan University, Australia. His current research interests lie in Chinese outbound tourism marketing, behaviours and other related aspects.

Liyu Yang is currently a postgraduate with the School of Tourism Sciences, Beijing International Studies University, China. Her research interests focus primarily on tourism big data, culture and tourism, tourism e-commerce, tourism marketing, hotel robots, traveller preferences and behaviour analysis.

Shaohua Yang is currently a Post-Doctoral Fellow from Graduate School of Business, Universiti Sains Malaysia, Malaysia. He holds Ph.D. in tourism marketing at Graduate School of Business, Universiti Sains Malaysia. His research interest lies in destination marketing, cross-cultural tourism, tourist behaviour, special interest tourism and self-congruency.

Elaine Yulan Zhang is a Lecturer in the Sino-British College at University of Shanghai for Science and Technology, Shanghai, China. Her interests also include digital marketing in tourism and hospitality, tourist decision-making and responsible management education.

Xiaodan Zhao is an independent researcher and a former Master's student in the Cultural Geography-Tourism and Planning Track at the Faculty of Spatial Sciences, University of Groningen, The Netherlands. Her research interests are tourist motivations and behaviour, tourism planning, destination marketing and tourism policy with a focus on Chinese tourists.

Lina Zhong is currently a Professor with the School of Tourism Sciences, Beijing International Studies University, China. Her research interests focus primarily on tourism data mining, culture and tourism, tourism marketing, hotel robots, tourism brand Image, hospitality management and planning.

Foreword

First, I would like to congratulate Dr. Jun Wen and Dr. Metin Kozak on having edited this high-quality book entitled *Chinese Outbound Tourist Behaviour: An International Perspective.* This book presents novel contributions regarding a rising international tourism market at a time when the industry has been hit hard by the COVID-19 pandemic. What distinguishes this resource from many others on similar topics is its use of industry language, as it brings together the views of tourism and travel industry experts from more than 11 countries. In probing numerous contemporary issues related to the Chinese outbound tourist market, the book's 17 chapters offer insight into how this market and Chinese tourists have evolved and how the country's tourism landscape is likely to expand in an era of uncertainty.

Dr. Jun Wen and I have known each other for over ten years and have collaborated on multiple projects involving Chinese outbound tourism. He is one of the most hardworking and productive scholars with whom I have had the pleasure of working. As a Chinese tourism researcher who has studied and worked overseas, his extensive professional experience in this area provided a solid basis for this book in many senses, from choosing the themes to inviting contributors. I have been pleased to witness Dr. Wen's growth since first meeting him as a Ph.D. student seeking research support.

Dr. Metin Kozak is an internationally recognized scholar who has offered great academic insight into tourism marketing over the years. I especially appreciate his mentorship of young scholars. Dr. Kozak's contributions to the tourism community will continue to be appreciated. In particular, I wish to express my gratitude for Dr. Kozak's mentorship of Dr. Wen in editing this book.

In an effort to offer a timely perspective on this valuable market along with suggestions to best harness the market's power, the two editors have selected thought-provoking papers from interdisciplinary studies in tourism and hospitality. I am confident that this book will be valuable to a wide audience; it is sure to serve research students, tourism practitioners and other relevant stakeholders well.

Tianyu Ying, Ph.D.
Zhejiang University, China

1 Chinese outbound tourist behaviour

An introduction

Jun Wen and Metin Kozak

Tourism's popularity as a research area is growing (Zhu et al., 2021a, 2021b). As an example, China has begun to attract increasing attention due to being one of the world's rapidly expanding source markets. Chinese outbound tourism is hence a topic of interest among industry stakeholders. These tourists accounted for nearly 155 million visitors in 2019—almost triple the number in 2010 (Textor, 2021; see Figure 1.1). Unfortunately, but not unexpectedly, COVID-19 caused this figure to decline to roughly 20.2 million in 2020. However, once the pandemic came under a degree of control, China's Labour Day holiday (1 May to 5 May 2020) implied some tourism recovery. Domestic visitors reached 115 million during the celebration, twice that of the Qingming Festival (otherwise known as Ancestors' Day or Chinese Memorial Day, an early April celebration). Tourism revenue was also five times higher—totalling 47.56 billion yuan—than during the Qingming Festival (Ministry of Culture and Tourism of China, 2020). About 90% of hotel enterprises in China had resumed operations by that time (China Hospitality Association, 2020). In fact, occupancy rates during the Labour Day holiday jumped more than twofold over the Qingming Festival (Tang, 2020). The Chinese thus appear to be willing to travel even though outbound tourism has yet to return to pre-pandemic levels; for instance, Australia has not begun to welcome international tourists as of this writing.

In past decades, China's outbound tourism market has risen faster than expected. The World Tourism Organization projected in 2004 that outbound Chinese tourists would total 100 million by 2020 (World Tourism Organization [WTO], 2004). That forecast was surpassed six years earlier than assumed—107.28 million Chinese outbound tourists travelled in 2014. According to China's Ministry of Culture and Tourism, Chinese citizens made 169.21 million outbound trips in 2019 alone (National Bureau of Statistics of China, 2020). Despite these promising statistics, pandemic-related travel restrictions have temporarily crippled outbound tourist flows as mentioned: only 20.23 million Chinese tourists were anticipated to travel overseas in 2020 (Figure 1.1). Yet, outbound travel among the Chinese is likely to pick up in the future—Ctrip and China Tourism Academy (2018) indicated that only 10% of China's population owns a passport and less than 10%

DOI: 10.4324/9781003121329-1

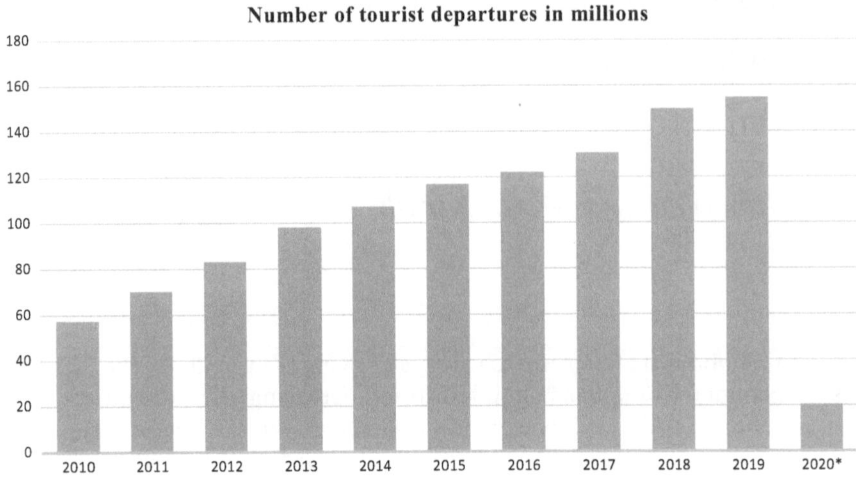

Figure 1.1 Number of outbound tourists from China between 2010 and 2019 with an estimate for 2020 (Textor, 2021)

are currently travelling abroad. These relatively small proportions reinforce the potential for ongoing growth in China's outbound tourism market post-COVID-19. Factors such as high travel demand and other trip purposes like overseas education should continue to propel this market.

Historical overview of the development of China's outbound tourism market

In order to have an overview of Chinese outbound tourist behaviour changes overtime since studies on Chinese outbound tourism began in the mid-1990s (Johnson et al., 2020), conducting a historical overview of the development of China's outbound tourism market is essential to lay the foundation (i.e., Arlt, 2016; Arlt & Burns, 2013; Bi et al., 2016; Lim, et al., 2020).

China's outbound tourism industry has undergone three developmental phases since 1983 (Lim et al., 2020). Prior to that year, outbound travel from China was strictly limited to business, education, or other official purposes (Lim et al., 2020). Leisure travel was typically considered wasteful (Arlt, 2016), causing international tourism to be exceedingly rare (Mak, 2013). The initial phase of modern outbound tourism in China occurred between 1983 and 1997. Travel during this period predominantly involved tourists who were visiting friends and relatives (VFR). These trips, which tourists first took to Hong Kong and Macao prior to being allowed to visit Southeast Asia, were mostly funded by relatives. VFR travel particularly enlarged China's tourism market to Hong Kong. This expansion was mostly due to the availability of package tours (e.g., for sightseeing) and two-way travel permits (Qu & Lam, 1997). Outbound tourism in mainland China then grew in 1991 following bilateral tourism agreements that enabled leisure tourists to travel on group tours coordinated by the Chinese Travel Service (Lim & Wang, 2008). Eligible countries included Malaysia, Singapore, and Thailand. Chinese residents thus had more opportunities to encounter life outside their own borders. Yet, of the 1.186 billion citizens in China in 1993, only 0.31% had travelled abroad (Arlt, 2016), underscoring this market's early growth potential.

The second phase of Chinese outbound tourism development spanned from 1997 to 2005. Throughout this period, the country's Approved Destination Status (ADS) scheme oversaw Chinese outbound tourism via bilateral travel agreements. China's Public Security Bureau and the China National Tourism Administration published "The temporary rules on Mainland Chinese outbound travel in self-financed mode" in 1997, confirming the establishment of outbound travel (Travel and Tourism Intelligence, 2003). ADS agreements were first signed with Australia and New Zealand in 1999 (Arlt, 2016; Lim & Wang, 2008). Additional agreements followed, including one with the European Union in 2004. Policies regarding travel to Hong Kong, Macau, and Taiwan were relaxed to foster tourism amid stronger cross-strait relations (Arlt, 2010). Outbound tourism in China ballooned from 3.7 million visitors in 1993 to 29 million in 2004 (Arlt, 2016). Other regions were inspired to try to capture these tourists as a result.

The third phase of Chinese outbound tourism began in 2005 and still applies today (Lim et al., 2020). The Chinese state has gradually reasserted regulatory control over outbound tourism through policies that accommodate the many Chinese tourists travelling overseas. These regulations simultaneously guarantee the government's power in protecting key policy objectives (Lim et al., 2020). The Chinese government at first promoted the development of its inbound travel market over domestic and outbound tourism (Wei & Wei, 2005). Yet, outbound tourism soon developed quickly under the guidance of President Xi Jinping, the first Chinese leader to offer such support since China's establishment in 1949. The Chinese government named tourism a "strategic pillar industry" in 2009 upon introducing detailed policies on the sustainability of outbound tourism.

Since then, outbound tourism has been one way that the country highlights its global standing (Tse, 2013). Mega events such as the 2008 Beijing Olympic Games and the Expo 2010 promoted China as a major tourist source market. The year 2012 represented another milestone in this market's development: China overtook the United States to become the largest outbound tourism source market in the world as measured by total departures and travel expenditure (Arlt, 2016). Chinese citizens spent $130 billion USD across roughly 150 million trips in 2018 (Tourism Research Institute, 2020). The market's prospects remain bright given that no more than 10% of the Chinese population currently hold passports. At present, Chinese outbound tourism can be segmented into various groups. These classifications naturally influence destination-related marketing and promotion as discussed below.

An overview of Chinese outbound tourism segmentation

Market segmentation can reveal travellers' features and help stakeholders devise tailored marketing approaches (Bloom, 2004). Segmentation has uncovered new and emerging markets in multiple destinations (Li et al., 2013). Several literature reviews to date have analysed the large body of research on Chinese outbound tourism (e.g., Law et al., 2016; Lojo et al., 2019); however, many studies have framed Chinese outbound tourists as a homogenous group. As a notable exception, Li et al. (2017) argued that China's long-haul outbound tourism market varies in terms of individuals' ages, motivations, preferences, travel experiences, travel arrangements, and so on. Quer and Peng (2021) delved into this assertion through a systematic review of studies on Chinese outbound tourism segmentation. They referred to 171 articles published between 1995 and 2020, focusing on specific segments of Mainland Chinese outbound tourists as outlined here.

The first identified Chinese outbound tourism segment consists of *shoppers* (Quer & Peng, 2021). Shopping is a major travel motivation among Chinese outbound tourists (e.g., Jin et al., 2020, 2021; Meng et al., 2019; Xu & McGehee, 2012), especially given Chinese citizens' growing consumption power. Scholars have started to pay particular attention to Chinese outbound tourists' luxury consumption, such as Chinese cigar tourists who travel to Cuba and other cigar-themed destinations (e.g., the Dominican Republic), to purchase authentic hand-rolled Cuban cigars and attend cigar-themed events such as cigar festivals (e.g., Ying et al., 2018). Quer and Peng (2021) also discussed the relationship between Chinese outbound tourists' shopping behaviour and cultural values; travellers' personal needs and shopping satisfaction were found to be influenced by product attitudes and service quality. Favourable shopping destinations for the Chinese include Hong Kong (Hung et al., 2020), Macao (Chang et al., 2017), Taiwan (Guo & Jiang, 2021), and North Korea (Li, Li et al., 2021). Quer and Peng (2021) further pointed out that Chinese outbound tourists' shopping satisfaction positively affects their travel-related attitudes as evidenced by destination loyalty and behavioural intentions.

Another Chinese outbound tourism segment consists of *students*, including in Australia (e.g., Li, Liu et al., 2021), the United States (e.g., Li, Lin, et al., 2021), New Zealand (e.g., Yang et al., 2021), and the United Kingdom (e.g., Song, 2021). With the emphasis on academic achievement in Chinese society, it comes as no surprise that some individuals opt to study abroad. Many choose to do so in pursuit of self-development (relational and individual), cross-cultural learning, high-quality education, language acquisition, and better job opportunities (Quer & Peng, 2021). At the same time, Chinese students also engage in leisure activities to balance their studies during daily life abroad. These students play an especially important role in overseas educational tourism for destinations such as Australia.

The third segment comprises *package tour and independent travellers*, who have been shown to exhibit unique travel preferences (Quer & Peng, 2021). Package tour travellers appreciate sights, cultural attractions, dining/eating, and tour guide services when visiting overseas, whereas independent travellers appreciate a destination's landscape, infrastructure and services, and travel facilitators. Independent tourists must plan trips themselves after performing sufficient research (Quer & Peng, 2021). These travellers seek trip-related information mainly from the media and their social networks. Interestingly, according to Ctrip (2017), independent travel among Chinese tourists has begun to take precedence over group arrangements (42% of trips are independent vs. 14% group tours). However, the coming years will likely see many first-time travellers take tours during their initial trips abroad. Experienced tourists visiting unusual or seemingly risky destinations may also take the same route before moving to self-organized travel. Research has similarly indicated that Chinese tourists visiting Israel, a volatile destination, prefer package tours in order to mitigate possible security concerns, cultural conflict, and other risks (Wen et al., 2019; Wen & Huang, 2019a, 2020). Still other scholars (e.g., Wen et al., 2020) have reported that Chinese senior tourists (aged 60–90) typically visit overseas destinations such as New Zealand, Israel, and Argentina through package tours.

Package tours will presumably remain popular within China's outbound tourism market. Certain tourist groups, such as seniors and people wishing to visit specific places (e.g., African countries, Israel, and other non-traditional destinations), may be especially likely to purchase package tours to reduce potential risks. Whereas younger generations of Chinese outbound tourists prefer to travel overseas independently for maximum freedom, inexperienced Chinese travellers who are unfamiliar with a destination and its language often choose to travel abroad on group tours. Most Chinese also remain influenced by their homeland's cultural traditions, which implicitly discourage travel to remote and unknown areas to limit possible risk (Xu et al., 2008). Ctrip (2017) reported that Chinese tourists, especially those from second-, third-, and fourth-tier cities and regions, still prefer group tours as well.

The fourth segment that Quer and Peng (2021) identified is *gamblers*. This form of special interest tourism is centred on Macao, the so-called "gambling capital" or "gaming mecca" of Asia. Researchers have explored special interest tourism among Chinese citizens with respect to cigar tourism (e.g., Wen & Huang, 2019b), wine tourism (e.g., Gu et al., 2017), adventure tourism (e.g., Jin et al., 2019), food tourism (e.g., Suntikul et al., 2020), and nature-oriented tourism (e.g., Fountain et al., 2010). These special interest studies reflect the diverse nature of tourist segments with unique interests and motivations for travelling to specific destinations overseas. More special interest tourist segments will likely be discerned in the near future and lead to a better overall understanding of the Chinese outbound tourist market. Meanwhile, though, a concrete operationalization is needed. To situate special interest tourism relative to late-modern leisure pursuits, McKercher and Chan (2005) posed the question "How special is special interest tourism?" in an effort to differentiate this type of tourism from "general" tourist activities. This query remains unresolved today. Wen and Wu (2020) more recently asked "How special are special interest tourists?" Having a clearer definition of and guidelines around special interest tourism can result in a firmer grasp on this market both in terms of Chinese outbound tourism and the global travel industry.

The fifth segment, *foodies*, suggests that Mainland Chinese tourists appreciate local food experiences (Quer & Peng, 2021). Physical surroundings and prestige are major concerns among Chinese travellers (Suntikul et al., 2020). Quer and Peng (2021) discovered that Chinese outbound foodies' local food experiences are largely based on available options (i.e., what to eat) as well as dishes' quality and presentation. These attributes can in turn affect tourists' destination loyalty in terms of visit intentions and willingness to recommend. Popular food tourism destinations from a Chinese perspective include Australia (e.g., Lai et al., 2020), Thailand (e.g., Promsivapallop & Kannaovakun, 2019), and Spain (e.g., Lin et al., 2020).

The last identified segment based on Quer and Peng's (2021) study is *millennials* (i.e., Generation Y). These tourists were born in the 1980s. Quer and Peng (2021) showed that Chinese millennial tourists have unique travel patterns characterised by self-development, cultural beliefs and family obligations, and preferences for less-common destinations. Key sub-segments of millennials include backpackers (e.g., Prayag et al., 2015) and working holiday makers (e.g., Zhu et al., 2020). However, as China is an ageing society with increasing emphasis on its senior population, it will be important to consider Chinese senior tourists in the near future. The following section provides an overview of three waves of Chinese outbound tourists since 1983 to contextualise anticipated trends in Chinese outbound tourist segmentation.

Three waves of Chinese outbound tourists

According to Bao et al. (2019), the first (older, unsophisticated), second (younger, more sophisticated), and third (upper-class middle-aged, sophisticated) waves of Chinese outbound tourists are distinct from one another. When the Chinese government started to allow citizens to travel overseas for leisure in the 1990s, the first wave of Chinese tourists (prior to 2005) favoured group travel packages—basically the sole leisure option under the ADS scheme (Arlt, 2016). These tourists typically travelled within Asia. Those who went outside the country frequented major cities and aimed to strictly control their accommodation and food expenditures (Arlt, 2013). Most Chinese tourists who went overseas wished to broaden their horizons by learning from Westerners (Xu et al., 2008). VFR and business tourists travelled abroad as well.

The second wave of Chinese outbound tourists consists of middle-aged individuals who are highly educated and affluent, often with extensive international travel experience (Arlt, 2016; Bao et al., 2019). These visitors, many of whom speak English, favour immersive destination experiences, and opt not to follow tour guides; they prefer instead to indulge in unique activities (Arlt, 2013). Spending power is a defining feature of this tourist wave. From 2005 to 2011, Chinese outbound tourists were the quickest growing market for luxury brand shopping. In fact, sales to Chinese customers were responsible for the largest proportion of luxury brands' earnings (Arlt, 2016).

Bao et al. (2019) adopted Sun and Wang's (2010) framework to profile the third wave of Chinese upper-class, middle-aged (40+ years old) tourists. Most countries, China included, are home to an ageing society. Travellers aged 50 and above are generally considered senior tourists (e.g., Hsu et al., 2007). According to the UNWTO, the over-60 segment will participate in more than 2 billion international trips by 2050 compared to 593 million throughout the last decade (Patterson, 2006). Of note, China's population is ageing more rapidly than that in any other country (Hsu et al., 2007): 400 million Chinese are expected to be age 60 or above by the mid-21st century (Mao & Song, 2004). While many tourist market segmentation studies have focused on younger travellers, secondary data have highlighted growth in the number of older Chinese tourists (Bao et al., 2019; Quer & Peng, 2021). Older generations of Chinese outbound tourists thus warrant attention when pondering future market trends. The third wave of Chinese outbound tourists is particularly susceptible to cultural and lifestyle changes as detailed in the ensuing section.

An overview of cultural changes and associated lifestyle changes

Culture greatly influences individuals' behaviour (Hofstede, 1980). Fulfilling Chinese travellers' demands and expectations requires an in-depth understanding of their cultural beliefs and values. Culture is an abstract concept, typically referring to standards shared by a group of people (Goodenough, 1971). Cultural values heavily inform Chinese consumers' behaviour (Tse & Hobson, 2008). Hofstede's (1980) national cultural framework is a common approach to explaining culture-based differences (Hsu et al., 2013). Leung (2008) pointed out that Chinese concepts such as face saving, *guanxi* (interpersonal connections), and harmony are affiliated with the broader cultural aspects of power distance and collectivism. These two dimensions distinguish Chinese culture from others (Hofstede, 1980).

Culture has affected Chinese outbound tourists across generations. Chinese individuals born between 1955 and 1975 tend to be more amenable to change (Egri & Ralston, 2004) and to possess modern values, such as individualism and secularity (Sun & Wang, 2010), compared to other generations—including those born during the "great leap forward" (people born before 1941) and the "cultural revolution" (people born between 1941 and 1956). These shifting cultural values likely shape travel motivations among China's middle-aged elite (e.g., *millennials*), including ego enhancement and self-esteem. For example, these groups enjoy independent trips more than group tours.

In particular, the Chinese assign great importance to the teachings of Confucius (551–479 BC). These lessons afford many citizens with valuable life guidance. Chinese culture is especially committed to upholding family obligations. These responsibilities encompass children's obligations to repay their parents for their sacrifices along with having respect for one's family hierarchy (Bao et al., 2019). As the sole generation to undergo the transformation from a controlled economy to a market economy—leading to the acquisition of property and wealth through one's own dedication within China's free-market economy—the middle-aged elite group holds unique attitudes about family obligations compared to their parents and children (Bao et al., 2019). China's one-child policy resulted in many single-child parents within this group. These cohort members have invested considerable effort and time in their children's educational success, work, and life quality while caring for their parents. Due to accumulated wealth and urban welfare schemes, these tourists are eligible for age-related support from the government or third-party organizations (e.g., insurance companies) on their parents' behalf. The elite group also has sufficient discretionary time and income for travel and leisure due to being empty nesters.

As China became integrated into the global community and permitted its citizens to travel internationally for business, education, and leisure, the elite have come to accept new cultures. Resultant attitudinal shifts have been enriched by exposure to Western business systems, expectations, organizational management methods, values, and so on. Compared with older generations, this cohort is more likely to consider "self-development" to be of paramount importance; this value proposition has influenced younger generations in particular (Sun & Wang, 2010). Associated changes in the elite group's cultural and social values have also informed their tourism participation.

Chinese senior outbound tourists are highly likely to be interested in travel, outside of unexpected crises such as COVID-19, due to cultural and lifestyle changes that have occurred throughout prior decades. Older Chinese generations have begun to accept tourism as a lifestyle component that enriches their physical and mental health. According to Xiong et al. (2021), Chinese adult children have started to pay closer attention to the support needed to help their elderly parents engage in outbound travel with confidence. This tourist segment will play a key role in China's outbound tourism market in the future. Travel motivations and other emerging topics are important to consider in this regard and will be discussed in the following section.

Travel motivations of three waves of Chinese outbound tourists

Motivation drives much of human behaviour. The concept of "tourist motivation" reflects why people wish to travel to a destination and/or take part in travel-related activities (Crompton, 1979). Scholars have begun to focus on Chinese tourists' motivations for travelling abroad. These visitors' prominent travel motivations can be divided based on the three tourist waves detailed previously.

Regarding the first wave of Chinese outbound tourists between 1983 and 2005, while Mainland Chinese tourists were initially permitted to travel to countries including Malaysia, Singapore, and Thailand, most related studies explored why they chose to visit Hong Kong. Chinese outbound tourists mainly travelled to visit friends and relatives in this region. The introduction of package tours later catalysed group travel (Qu & Lam, 1997). Since 1997, Hong Kong has remained the most frequently studied destination from the perspective of Chinese travellers. Affiliated travel motivations include the "push factors" of *enhancement of human relationships*, *knowledge*, *novelty*, *prestige*, and *relaxation* along with the "pull factors" of *accessibility*, *cultural links*, *expenditure*, *hi-tech image*, *service attitude and quality*, and *sightseeing variety* (Hanqin & Lam, 1999).

The second wave of Chinese outbound tourists emerged in 2005. A growing body of literature has since examined these tourists' motivations for visiting overseas destinations. Popular locations for the Chinese market include New Zealand, thanks in large part to its breathtaking natural landscapes, artisanal souvenirs, and museums (Fountain et al., 2010) as well as relaxation and potential career and educational opportunities (Ryan & Zhang, 2007). Traditionally, popular tourism destinations for Chinese travellers include South Korea, Canada, and Australia. Other seemingly unusual locations, such as North Korea and Israel, have been examined as well. For example, international tourism to Israel grew steadily in 2016: the country saw 2.9 million tourist arrivals (a 4% increase over 2015).

Of these visitors, 86,000 were Chinese, leading to a 30% jump compared with 2015 (Tourism Ministry of Israel, 2017). This expansion reflects another new trend in Chinese tourists choosing destinations plagued by crises (e.g., Wen et al., 2019). In 2017, tourist arrivals to Israel jumped 25% to 3.6 million; over 113,000 were Chinese, constituting Israel's quickest-growing tourist source market. Wen and Huang (2020) examined Chinese outbound tourists' reasons for going to Israel based on means-end theory and a laddering method. They ultimately discerned six means-end patterns that captured travellers' motivations. Findings indicated key destination attributes driving Chinese tourists to visit: geographical location, history, local culture, local people (Jews), religions, and well-known places. These characteristics represented the three end values of life *enrichment*, *self-esteem*, and *achievement*.

Wen et al. (2019) explored relationships among Chinese cultural values and Chinese tourists' motivations for visiting Israel. Results showed that modern values influenced travellers' motivations for business development and self-fulfilment; the motivation for sightseeing was tied to life enrichment and quality. Wen and Huang (2019a) segmented the same tourist group through a motivation-based framework of Chinese cultural values, identifying adventure-driven mass tourists, enthusiastic tourists, mass tourists, and sightseeing and horizon-enhancing tourists. Given the growing popularity of Chinese outbound tourism to unusual and risky destinations, more studies are needed to explore travellers' underlying motivations and behaviour across these types of locales.

As Chinese tourists aged 40 and above will eventually enter an elderly phase of life, the middle-aged elite group will likely come to embrace forms of travel such as family tourism. This form of tourism has been fuelled by efforts to reinforce one's family values, connect with relatives, and strengthen family identity (Wu et al., 2021). Family travel incorporates family members' unique needs (Gram, 2005), enabling the Chinese middle-aged elite to coordinate travel activities that unite their family. Initial research on family travel from a Chinese point of view has documented Chinese adult children's reasons for travelling with their parents: commemoration and celebration, compensation and reward, dream fulfilment, escape and relaxation, expressions of filial piety, family togetherness, health, novelty and knowledge, and pleasure-seeking (Wang et al., 2018).

These motivations can be divided into family-oriented, parent-oriented, and self-oriented facets, classifying distinct types of adult children–senior parent travellers and capturing familial norms in China's ageing society. Undoubtedly, more research is needed to provide a full understanding of this segment from different angles in the context of outbound tourism. For instance, it is important to identify Chinese senior outbound tourists' perceived constraints to eliminate barriers to engaging in tourism (e.g., Wen et al., 2020). It will also be useful to uncover this segment's negotiation strategies when partaking in outbound tourism and their psychological status during the decision-making process (e.g., Wen et al., 2020). The roles of smart services in Chinese senior tourists' trips offer one example of how older tourists will need to adapt to a changing tourism environment (e.g., Jiang & Wen, 2020).

Synopsis of contributing chapters to this book

Based on a review of the main stages of Chinese outbound tourism development, three prominent waves of Chinese tourists, and relevant travel motivations, the evidence suggests that these tourists' profiles have changed substantially throughout the past four decades. Despite ongoing academic research into this tourism market and visitors' travel behaviour, scholars should continue to unearth new insights around these tourists.

Compared with other countries, China's foray into international tourism is relatively recent with respect to tourist generation and reception.

The country's unique culture, nature, and architecture offer an array of tourism attractions, positioning China as an enticing destination. China is home to the largest population in the world and boasts the second largest proportion of outbound travellers and the highest visitor spending in international tourism. Current projections suggest that China will continue to grow to become the top tourism market and destination within the next few years. This book seeks to capture China's potential as a travel hub to promote its success in tourism. China, as a young entrant onto the international tourism scene, deserves careful exploration thanks to its remarkable cultural features.

With the contribution of 35 authors representing 11 countries, this volume of 17 chapters assumes an international perspective. This collection traces the evolution of China's tourism market, explores Chinese outbound tourists' behaviour, and considers how the country's tourism landscape will expand. Specifically, the book offers a vivid profile of Chinese tourists and the characteristics differentiating them from other market segments. We present an overview of Chinese tourist segments and travel-related concerns to paint a clear picture of the market's status. Armed with knowledge of Chinese outbound tourists' behavioural characteristics, the supply and demand sides of the country's tourism industry will receive information necessary for their respective development. On the supply side, practitioners can implement effective strategies to meet and exceed consumers' needs and expectations; on the demand side, satisfying and engaging tourism experiences will enhance visitors' quality of life and general trips. The book's closing chapters address the future of Chinese tourism. Ultimately, this collection is intended to provide industry stakeholders an up-to-date view on this valuable market along with suggestions to best harness the market's power.

This book contains four parts. This section provides a synopsis of each chapter. The first part, consisting of four chapters, maps the historical development of Chinese outbound tourism. The first chapter provides an overview of related history, development, and politics. It focuses on policies and regulations that have shaped Chinese outbound tourism, including the Travel Agency Act, the ADS Scheme, and the Chinese Tourism Law. The second chapter explores the behavioural characteristics of Chinese outbound tourists, namely their motivations for travelling abroad. The third chapter looks at factors responsible for the growth, size, and politics of outbound tourism, with a note about the pandemic's impacts. The last chapter provides an overview of the evolution of China's outbound tourism as well as an analysis of emerging market trends and travel patterns. All chapters end with a reflection on the past and future development of Chinese outbound tourism.

The second part, with eight chapters, reports on the travel experiences of Chinese outbound tourists in various countries. The first chapter examines major motivations driving Chinese tourists to visit Italy, including cultural heritage, authenticity seeking, and luxury travel. The second chapter recommends marketing strategies to attract Chinese tourists to Malaysia through wellness tourism. As an empirical study, the third chapter suggests practical strategies to help local tourism shareholders expand their market structure by entering the Chinese tourism market. The next chapter identifies weaknesses that (in)directly influence Chinese tourists' perceptions of service quality in New Zealand. The fifth chapter categorizes Chinese tourists' behaviour on a group tour to Turkey. The sixth chapter emphasizes food tourism in shaping destination image and how this form of tourism can enhance Chinese tourists' behavioural intentions. The last two chapters reflect on Chinese tourists' motivations and perceptions—from experiencing an authentic Dutch identity at tulip fields in the Netherlands to how the pandemic has hindered Chinese visitation to Australia, leading to a sharp decline in tourism income.

The third part includes three chapters on Chinese outbound tourists' shopping and health. The first chapter analyses these travellers' shopping behaviour and how it can vary by country. The next chapter explores underlying meanings of luxury travel, resulting in a set of six themes: wealth, facilities, service quality, destination experience, authentic, slow and pleasure-seeking, and contrast between travel experience and daily life. This part closes with a chapter on the effects of the pandemic on Chinese citizens' intentions to visit Turkey for medical reasons, which represents a growing market in international tourism.

The final part is a single chapter that revisits how Chinese outbound tourism is likely to alter the tourism industry. The chapter suggests that any shifts in the Chinese market, as the world's largest tourist population, may present global supply-side challenges and advantages. The demand side is expected to respond to such changes by travelling either in smaller groups or alone.

As an end note, we hope that this book will offer readers greater knowledge about the past, present, and future of the Chinese outbound tourism market.

References

Arlt, W. G. (2013). The second wave of Chinese outbound tourism. *Tourism Planning & Development, 10*(2), 126–133. doi:10.1080/21568316.2013.800350

Arlt, W. G. (2016). *China's outbound tourism: History, current development, and outlook*. Oakville: Apple Academic Press.

Arlt, W. G. (Ed.). (2010). *China Tourism Academy: Annual report Chinese outbound tourism development 2009/2010* (English edition). Profil Verlag: München.

Arlt, W. G., & Burns, P. (2013). Chinese outbound tourism. *Tourism Planning & Development, 10*(2), 123–125.

Bao, J., Jin, X., & Weaver, D. (2019). Profiling the elite middle-age Chinese outbound travellers: A 3rd wave?. *Current Issues in Tourism, 22*(5), 561–574. doi:10.1080/13683500.2018.1449817

Bi, L., Vanneste, D., & van der Borg, J. (2016). Tourism development in China since 1949: A regime perspective. *International Journal of Tourism Policy, 6*(3–4), 321–340.

Bloom, J. Z. (2004). Tourist market segmentation with linear and non-linear techniques. *Tourism Management, 25*(6), 723–733. doi:10.1016/j.tourman.2003.07.004

Chang, T.-Z. (Donald), Kong, W. H., & Chen, S.-J. (2017). The moderating role of environmental concerns in travel behavior: An exploratory study. *Journal of China Tourism Research, 13*(2), 193–210. doi:10.1080/19388160.2017.1353471

China Hospitality Association. (2020). *China Hotel Association Research Institute released: March report on the current status of China's accommodation industry under the new crown epidemic.* https://mp.weixin.qq.com/s?__biz=MjM5MTIyMDYyMw==&mid=2652013339&idx=1&sn=ca1c913caa74903f9ecb0fed00bf4a86&chksm=bd5ef5fb8a297cedfd9e312caa4002a679ad91c0cd05c1dfdb06b4a6f5fc12069af32c784eb7&token=438481508&lang=zh_CN#rd

Crompton, J. L. (1979). Motivations for pleasure vacation. *Annals of Tourism Research, 6*(4), 408–424. doi:10.1016/0160–7383(79)90004-5

Ctrip. (2017). *2017 China outbound tourism travel report.* Retrieved June 30, 2021, from https://www.chinatravelnews.com/images/201802/fc66f776a9111201.pdf

Ctrip & China Tourism Academy. (2018). *2017 China outbound tourism travel report.* Ctrip and China Tourism Academy.

Egri, C., & Ralston, D. (2004). Generation cohorts of personal values: A comparison of China and the United States. *Organization Science, 15*(2), 210–220. doi:10.1287/orsc.1030.0048

Fountain, J., Espiner, S., & Xie, X. (2010). A cultural framing of nature: Chinese tourists' motivations for, expectations of, and satisfaction with, their New Zealand tourist experience. *Tourism Review International, 14*, 71–83. doi:10.3727/154427211X13044361606298

Goodenough, W. (1971). *Culture, language, and society.* Reading, MA: Addison-Wesley.

Gram, M. (2005). Family holidays. A qualitative analysis of family holiday experiences. *Scandinavian Journal of Hospitality and Tourism, 5*(1), 2–22. doi:10.1080/15022250510014255

Gu, Q., Qiu Zhang, H., King, B., & Huang, S. (Sam). (2017). Wine tourism involvement: A segmentation of Chinese tourists. *Journal of Travel & Tourism Marketing, 35*(5), 633–648. doi:10.1080/10548408.2017.1401031

Guo, L., & Jiang, F. (2021). Is cross border tourism an effective means to achieve political goals? Evidence from Mainland China and Taiwan. *Applied Economics Letters*, 1–7. doi:10.1080/13504851.2021.1899116

Hanqin, Z. Q., & Lam, T. (1999). An analysis of Mainland Chinese visitors' motivations to visit Hong Kong. *Tourism Management, 20*(5), 587–594. doi:10.1016/s0261–5177(99)00028-x

Hofstede, G. (1980). Culture and organizations. *International Studies of Management & Organization, 10*(4), 15–41. doi:10.1080/00208825.1980.11656300

Hsu, C. H. C., Cai, L. A., & Wong, K. K. F. (2007). A model of senior tourism motivations—Anecdotes from Beijing and Shanghai. *Tourism Management, 28*(5), 1262–1273. doi:10.1016/j.tourman.2006.09.015

Hsu, S.-Y., Woodside, A. G., & Marshall, R. (2013). Critical tests of multiple theories of cultures' consequences: Comparing the usefulness of models by Hofstede, Inglehart and Baker, Schwartz, Steenkamp, as well as GDP and distance for explaining overseas tourism behavior. *Journal of Travel Research, 52*(6), 679–704. doi:10.1177/0047287512475218

Hung, K., Qiu Zhang, H., Guillet, B. D., & Wang, L. (2020). China watching: Luxury consumption and its implications. *Journal of Travel & Tourism Marketing, 37*(5), 577–592. doi:10.1080/10548408.2018.1525470

Jiang, Y., & Wen, J. (2020). Effects of COVID-19 on hotel marketing and management: A perspective article. *International Journal of Contemporary Hospitality Management, 32*(8), 2563–2573. doi:10.1108/ijchm-03–2020–0237

Jin, H., Moscardo, G., & Murphy, L. (2020). Exploring Chinese outbound tourist shopping: A social practice framework. *Journal of Travel Research, 59*(1), 156–172. doi:10.1177/0047287519826303

Jin, H., Moscardo, G., & Murphy, L. (2021). Unraveling the mechanisms behind Chinese outbound tourist shopping: A social practice perspective. *Journal of Hospitality & Tourism Research, 45*(4), 629–651. doi:10.1177/1096348020956690

Jin, X., Xiang, Y., Weber, K., & Liu, Y. (2019). Motivation and involvement in adventure tourism activities: A Chinese tourists' perspective. *Asia Pacific Journal of Tourism Research, 24*(11), 1066–1078. doi:10.1080/10941665.2019.1666152

Lai, M. Y., Wang, Y., & Khoo-Lattimore, C. (2020). Do food image and food neophobia affect tourist intention to visit a destination? The case of Australia. *Journal of Travel Research, 59*(5), 928–949. doi:10.1177/0047287519867144

Law, R., Sun, S., Fong, D. K. C., Fong, L. H. N., & Fu, H. (2016). A systematic review of China's outbound tourism research. *International Journal of Contemporary Hospitality Management, 28*(12), 2654–2674. doi:10.1108/ijchm-06–2015–0323

Leung, K. (2008). Chinese culture, modernization, and international business. *International Business Review, 17*(2), 184–187. doi:10.1016/j.ibusrev.2008.02.009

Li, C., McCabe, S., & Chen, Y. (2017). Destination choice of Chinese long-haul outbound tourists and market segmentation. *Journal of China Tourism Research, 13*(3), 298–315. doi:10.1080/19388160.2017.1398118

Li, R., Lin, Y., Miller, Z., Su, C., Pan, B., Taff, B., & Graefe, A. (2021). An exploratory study on Chinese tourists' visitation to a U.S. National Park. *Tourism Recreation Research*, 1–15. doi: 10.1080/02508281.2021.1889792

Li, T. (Tina), Liu, F., & Soutar, G. N. (2021). Experiences and value perceptions of an ecotourism trip—an empirical study of outbound Chinese tourists. *Tourism Recreation Research, 46*(3), 333–344. doi:10.1080/02508281.2020.1804736

Li, X. (Robert), Meng, F., Uysal, M., & Mihalik, B. (2013). Understanding China's long-haul outbound travel market: An overlapped segmentation approach. *Journal of Business Research, 66*(6), 786–793. doi:10.1016/j.jbusres.2011.09.019

Li, Z., Li, F. (Sam), & Ryan, C. (2021). Perceiving North Korea through Chinese tourists' eyes. *Tourism Review, 76*(1), 150–163. doi:10.1108/tr-12–2019–0488

Lim, C., & Wang, Y. (2008). China's post-1978 experience in outbound tourism. *Mathematics and Computers in Simulation*, 78(2–3), 450–458. doi:10.1016/j.matcom.2008.01.033

Lim, D. J., Ferguson, V. A., & Bishop, R. (2020). Chinese outbound tourism as an instrument of economic statecraft. *Journal of Contemporary China, 29*(126), 916–933. doi:10.1080/10670564.2020.1744390

Lin, J., Julve, J., Xu, H., & Cui, Q. (2020). Food habits and tourist food consumption: An exploratory study on dining behaviours of Chinese outbound tourists in Spain. *Journal of Policy Research in Tourism, Leisure and Events, 12*(1), 82–99. doi:10.1080/19407963.2018.1506199

Lojo, A., Li, M., & Cànoves, G. (2019). Co-authorship networks and thematic development in Chinese outbound tourism research. *Journal of China Tourism Research, 15*(3), 295–319. doi:10.1080/19388160.2018.1512433

Mak, B. (2013). The influence of political ideology on the outbound tourism in China. *Journal of China Tourism Research*, *9*(1), 1–26. doi:10.1080/19388160.2013.756764

Mao, Y., & Song, G. (2004). The aging issue and the initial research on seniors' tourism marketing strategies. *Journal of West Anhui University*, *21*(1), 92–94.

McKercher, B., & Chan, A. (2005). How special is special interest tourism?. *Journal of Travel Research*, *44*(1), 21–31. doi:10.1177/0047287505276588

Meng, F., Zhang, P., Li, H., & So, K. K. F. (2019). Modeling precursors of impulsive tourist shopping behavior: Evidence from long-haul Chinese outbound tourists. *International Journal of Tourism Research*, *21*(3), 344–358. doi:10.1002/jtr.2266

Ministry of Culture and Tourism of China. (2020). *May Day holiday: 115 million people travel around nationwide*. https://www.mct.gov.cn/whzx/whyw/202005/t20200505_852977.htm

National Bureau of Statistics of China. (2020, February 28). Statistical Communiqué of the People's Republic of China on the 2019 national economic and social development. http://www.stats.gov.cn/english/PressRelease/202002/t20200228_1728917.html

Patterson, I. (2006). *Growing older: Tourism and leisure behaviour of older adults*. Oxfordshire and Cambridge, MA: CABI.

Prayag, G., Disegna, M., Cohen, S. A., & Yan, H. (2015). Segmenting markets by bagged clustering: Young Chinese travelers to Western Europe. *Journal of Travel Research*, *54*(2), 234–250. doi:10.1177/0047287513514299

Promsivapallop, P., & Kannaovakun, P. (2019). Destination food image dimensions and their effects on food preference and consumption. *Journal of Destination Marketing & Management*, *11*, 89–100. doi:10.1016/j.jdmm.2018.12.003

Qu, H., & Lam, S. (1997). A travel demand model for Mainland Chinese tourists to Hong Kong. *Tourism Management*, *18*(8), 593–597. doi:10.1016/s0261-5177(97)00084-8

Quer, D., & Peng, J. (2021). Chinese outbound tourism segmentation: A systematic review and research agenda. *Journal of China Tourism Research*. doi:10.1080/19388160.2021.1966563

Ryan, C., & Zhang, Z. (2007). Chinese students: Holiday behaviours in New Zealand. *Journal of Vacation Marketing*, *13*(2), 91–105. doi:10.1177/1356766707074734

Song, H. (2021). Linking images to information sources: An examination of potential Chinese tourists' multi-sensory images of the United Kingdom. *Journal of China Tourism Research*. doi:10.1080/19388160.2021.1877860

Sun, J., & Wang, X. (2010). Value differences between generations in China: A study in Shanghai. *Journal of Youth Studies*, *13*(1), 65–81. doi:10.1080/13676260903173462

Suntikul, W., Pratt, S., & Chong, Y. W. J. (2020). Factors that Influence Chinese Outbound Tourists' Intention to Consume Local Food. *Journal of China Tourism Research*, *16*(2), 230–247. doi:10.1080/19388160.2019.1625841

Tang, Y. (2020). The number of tourists staying at the hotel during the 'May Day' holiday has tripled compared with Qingming. *The Paper*. Retrieved June 30, 2021, from https://www.thepaper.cn/newsDetail_forward_7278028

Textor, C. (2021, August 18). Number of outbound tourists departing from China from 2010 to 2019 with an estimate for 2020. *Statista*. Retrieved October 1, 2021, from https://www.statista.com/statistics/1068495/china-number-of-outbound-tourist-number/

Tourism Ministry of Israel. (2017). *Domestic and foreign tourists number of overnights 2017.* Retrieved June 30, 2021, from https://info.goisrael.com/en/domestic-and-foreign-tourists-number-of-overnights-2017

Tourism Research Institute. (2020). *China outbound tourism development annual report 2019.* http://www.pinchain.com/article/198626

Travel and Tourism Intelligence. (2003). China outbound. *Travel and Tourism Analyst, 1*, 1–32.

Tse, T. S. M. (2013). Chinese outbound tourism as a form of diplomacy. *Tourism Planning & Development, 10*(2), 149–158. doi:10.1080/21568316.2013.783738

Tse, T. S. M., & Hobson, J. (2008). The forces shaping China's outbound tourism. *Journal of China Tourism Research, 4*(2), 136–155. doi:10.1080/19388160802279459

Wang, W., Yi, L., Wu, M.-Y., Pearce, P. L., & Huang, S. (Sam). (2018). Examining Chinese adult children's motivations for traveling with their parents. *Tourism Management, 69*, 422–433. doi:10.1016/j.tourman.2018.06.024

Wei, F., & Wei, C. (2005). A study on the trends in China outbound tourism market. In *Proceedings of APTA Conference South Korea.*

Wen, J., & Huang, S. (2019a). Chinese tourists visiting volatile destinations: Integrating cultural values into motivation-based segmentation. *Journal of China Tourism Research, 15*(4), 520–540. doi:10.1080/19388160.2019.1589610

Wen, J., & Huang, S. (2019b). The effects of push and pull travel motivations, personal values, and destination familiarity on tourist loyalty: A study of Chinese cigar tourists to Cuba. *Asia Pacific Journal of Tourism Research, 24*(8), 805–821. doi:10.1080/10941665.2019.1635504

Wen, J., & Huang, S. (2020). Chinese tourists' motivations of visiting a highly volatile destination: A means-end approach. *Tourism Recreation Research, 45*(1), 80–93. doi:10.1080/02508281.2019.1637078

Wen, J., Huang, S., & Goh, E. (2020). Effects of perceived constraints and negotiation on learned helplessness: A study of Chinese senior outbound tourists. *Tourism Management, 78*, 104059. doi:10.1016/j.tourman.2019.104059

Wen, J., Huang, S. (Sam), & Ying, T. (2019). Relationships between Chinese cultural values and tourist motivations: A study of Chinese tourists visiting Israel. *Journal of Destination Marketing & Management, 14*, 100367. doi:10.1016/j.jdmm.2019.100367

Wen, J., & Wu, M.-Y. (2020). How special is special interest tourism—and how special are special interest tourists? A perspective article in a Chinese context. *Current Issues in Tourism, 23*(16), 1968–1972. doi:10.1080/13683500.2020.1750575

World Tourism Organization. (2004). *WTO in Asia and the Pacific 2004.* UNWTO. https://www.e-unwto.org/doi/book/10.18111/9789284408184

Wu, W., Wu, M.-Y., Yi, L., & Lehto, X. (2021). Measuring Chinese adolescents' learning outcomes in family travel: A scale development approach. *Journal of Destination Marketing & Management, 20*, 100617. doi:10.1016/j.jdmm.2021.100617

Xiong, Q., Zhang, Q., Yang, Y., & Li, Z. (2021). Perceived barriers faced by Chinese adult children in social support of their elderly parents' outbound travels. *Tourism Management Perspectives, 38*, 100827. doi:10.1016/j.tmp.2021.100827

Xu, H., Ding, P., & Packer, J. (2008). Tourism research in China: Understanding the unique cultural contexts and complexities. *Current Issues in Tourism, 11*(6), 473–491. doi:10.1080/13683500802475737

Xu, Y., & McGehee, N. G. (2012). Shopping behavior of Chinese tourists visiting the United States: Letting the shoppers do the talking. *Tourism Management*, *33*(2), 427–430. doi:10.1016/j.tourman.2011.05.003

Yang, S., Isa, S., Ramayah, T., Wen, J., & Goh, E. (2021). Developing an extended model of self-congruity to predict Chinese tourists' revisit intentions to New Zealand: The moderating role of gender. *Asia Pacific Journal of Marketing and Logistics*. doi:10.1108/APJML-05–2021–0346

Ying, T., Wei, W., Wen, J., Wang, L., & Ye, S. (2018). Chinese cigar tourists to Cuba: A motivation-based segmentation. *Journal of Destination Marketing & Management*, *10*, 112–121. doi:10.1016/j.jdmm.2018.07.003

Zhu, H., Tucker, H., Duncan, T., & Zhang, J. (2020). Young Chinese becoming working holiday makers in New Zealand: A thematic analysis of WHMs' motivations. *Journal of China Tourism Research*, *16*(3), 406–423. doi:10.1080/19388160.2019.1650154

Zhu, J. (Jason), Airey, D., & Siriphon, A. (2021a). Chinese outbound tourism: An alternative modernity perspective. *Annals of Tourism Research*, *87*, 103152. doi:10.1016/j.annals.2021.103152

Zhu, J., Airey, D., & Siriphon, A. (2021b). Chinese outbound tourists as international consumer in Northern Thailand—A dynamic mobility perspective. *Journal of Consumer Culture*, 1–19. doi:10.1177/1469540521994318

Part I

Historical development of Chinese outbound tourism

2 History, development and politics of China's outbound tourism

Matias Thuen Jørgensen and Anders Ellemann Kristensen

Introduction

Tsang and Hsu (2011) argue that China's tourism and hospitality industry has undergone three major stages in order to reach its current state. The first stage started with the founding of the People's Republic of China in 1949. Here, tourism served mainly a political purpose. Its goal was to promote the country's accomplishments and to build or improve international relations (Huang, 2010; Tsang & Hsu, 2011). In the second stage, the main purpose and nature of tourism was changed from strictly political to economical aspirations. This period started with implementation of economic reforms in 1978 and ended with China officially joining the UNWTO in the mid-1980s. The third stage began in 1986, with the inclusion of tourism as an important factor for social and economic development in the Seventh Five-year National Plan. This event was important because the purpose of tourism was officially changed from a dual focus on economics and politics to a more specific focus on the economic aspects rather than the political (Zhang et al., 1999).

Although these phases illustrate that tourism in China is as old as The Peoples Republic, outbound tourism from China only emerged in the third stage. The first outbound tourism from China began in 1983, shortly after the establishment of the China National Tourism Administration (CNTA), as visits to friends and relatives in Hong Kong and Macau were allowed (Qu & Lam, 1997). However, since outbound tourism from China was still heavily regulated, and since the number of people with disposable income was still limited, it was not until the late 1990s that outbound tourism became anything resembling what it is today (Kristensen, 2020).

The CNTA "... plays the role of both tourist industry authority auditing the operation of tour operators, hotels and tour education institutions in China and destination image promoter via its overseas branches" (Ma et al., 2003, p. 461). A unique attribute of the CNTA was and is that its mission is not only to attract tourism to China, as is often the case for national destination marketing/management organisations, but also to set guidelines for outbound tourism from the country through regulation of tour operators,

DOI: 10.4324/9781003121329-3

hotels and other actors in the industry. This is an indication of the important role that policymaking has had in influencing China's outbound tourism.

A number of important political changes have shaped China's outbound tourism into what it is today. This chapter will use three policies, which have marked such change, as cornerstones to outline and discuss the development of China's outbound tourism, namely the 'Travel Agency Act', the 'Approved Destination Status' scheme and the 'Chinese Tourism Law'. We end the paper with a discussion of the past and future development of the China's outbound tourism market.

The Travel Agency Act – How it halted competition and shaped the OTAs

The 'Provisional Regulations on the Administration of Travel', better known as the 'Travel Agency Act', promulgated in 1985, was China's first administrative tourism legislation. It paved the way for the first outbound tours from Greater China around 1990. Trips were only allowed to Malaysia, Singapore, the Philippines, and Thailand, and only for family visits, although many of the trips clearly had a leisure purpose (Lim et al., 2020). In 1995, the list of countries was expanded (Guo et al., 2007), and the Act was revised and retitled to 'Regulations on the Administration of Travel Agencies' in 1996. The act was supported by simplified procedures to obtain passports, international currency and visas.

Jørgensen et al. (2020, p. 20) have summarised the Act as follows:

> This act defined two types of travel agency: International travel agencies, which could operate inbound, outbound and domestic tourism; and domestic travel agencies, which were confined to the domestic tourism arena (Huang, 2010). Further revisions to the act came about in 2001 following China's accession to the World Trade Organization (2000). One of the most important changes was the permission for foreign travel agencies to establish joint-venture travel agencies in Beijing, Shanghai, Guangzhou and Xian under conditions which were progressively softened and finally eliminated in 2005 (Zhang, 2004). Only one condition remained – joint venture companies were still not allowed to engage in Chinese outbound tourism.

It is notable that even if regulations in the early 2000s allowed for foreign competition, only a few licences to offer outbound services for international companies have been issued thus far, and that the first licence was issued as late as May 2011 (Lim et al., 2020). It is evident that as travel quickly became digitalised, in the early 2000s, the act could not accommodate a rapid changing online travel environment or foreign investments therein. This instead forced a number of Chinese online travel agencies (OTAs) to work in a regulatory grey area using a variable interest enterprise (VIE) business model (Gillis, 2011).

The need for foreign investors and foreign capital resulted in number of Chinese OTAs listed on American stock exchanges. Ctrip.com's own F1 filing to the SEC shows that the purpose of the VIE was to the possibility to adopt a business model that allowed for foreign ownership (Ctrip, 2003, p. 9).

The effects of the act are still felt today, as some large Chinese OTAs, including Ctrip (now Trip.com) and Tuniu, are still listed on the NASDAQ. Regulatory solutions for foreign investments directly into Chinese OTAs have yet to be offered. This combination of regulatory restrictions on foreign investors has created a closed environment within the Chinese travel industry, where little competition from foreign investors exists. The travel patterns of Chinese travellers are changing from being primarily group package tourists who arrange their travel through brick-and-mortar agencies towards online tour booking and semi-independent travellers, who purchase some trip elements through OTAs. This development transfers increasing power to these OTAs in the distribution of China outbound tourists.

Approved Destination Status (ADS) – How outbound tourism became soft power

The first self-supported outbound tourism from China was possible via the ADS Scheme, which was initiated in 1997. The ADS scheme lists companies authorised as able to handle bookings and visas for tour groups. The ADS agreement only permits group travel between China and another country if they both sign a memorandum of understanding. In effect, this regulates both agencies at the destinations and Chinese agencies sending tourists abroad (Lim et al., 2020). The purpose of the ADS scheme is threefold: (1) it manages the outflows of tourists from China to different countries, also addressing receiving destination countries' concerns of possible illegal immigration; (2) it helps to alleviate otherwise difficult visa processes and (3) it helps to control domestic travel agencies and international tour operators (Dai et al., 2017; Lim et al., 2020; Tse, 2011).

Through ADS, some Chinese operators were chosen to arrange, market and sell trips to specific destinations in collaboration with approved international companies that provide inbound services (Pan & Laws, 2003). A country can only receive ADS if they abide by the following guidelines. The country must:

> (1) generate outbound tourists to China; (2) have favourable political relations with China; (3) have attractive and suitable tourism resources and facilities for Chinese travellers; (4) guarantee safety and freedom from discrimination for Chinese travellers; (5) be easily accessible; (6) have a balance of payments in terms of arrivals and expenditures with China; (7) have a reciprocal increase in the share of tourists from the country to China, and from China to the country.
>
> (Guo et al., 2007)

In the 2000s, the ADS system was viewed positively by the countries receiving Chinese travellers as it was a means to safeguard against illegal immigration, while giving the receiving countries channels to handle larger tourist groups (Arita et al., 2012). ADS was first given to a few neighbouring countries in Asia and has since been widened to include countries all over the world. In 2018, 130 countries and territories had received ADS (Lim et al., 2020). In 2013, the UNWTO predicted that the ADS would be relaxed gradually and consequently phased out in the future (ETC & UNWTO, 2013). While it has since been progressively loosened and other countries have been added to the list, there are signs that it may not be phased out as predicted.

A reason for this could be that outbound tourism is being utilised as a soft power tool by the Chinese government, among other things through the ADS scheme (Dai, 2011; Tse, 2013; Xu et al., 2020). Dai et al. (2017), for example, argued for Chinese outbound tourism to be utilised as a soft power tool:

> We should take advantage of the rapid development of the outbound tourism to enhance the Chinese voice in international affairs, especially those of organizations, policies, standards and other aspects. In the process of ADS negotiation, implementation, and assessment, it is possible to guide the destination countries to gradually open television access to Chinese satellite channels, and a more liberal import policy of Chinese books, magazines and newspapers. Additionally, an improvement in the Chinese language hosting environment can be an important requirement in the opening and implementation of Chinese outbound tourism.
>
> (p. 257)

In addition to such calls, there are also arguments and empirical evidence indicating that the ADS scheme is already being utilised as a soft power tool. Lim et al. (2020) noted that:

> It has been widely reported that Beijing has placed formal and informal restrictions on outbound tourism to the Philippines, Japan, Taiwan, and South Korea as punishment for pursuing policies at odds with Chinese interests. [...] Some analysts assert that Beijing has 'weaponized' tourism by discouraging travel to the United States. (p. 1) [...] Negotiating new ADS agreements became a tool through which Beijing could confer rewards to potential destination states.
>
> (p. 8)

Three concrete examples that support this include: (1) A 15-month diplomatic standoff between China and Turkey, which began when Turkey refused a Chinese carrier passage through the Bosporus strait in 2000 and ended when China offered a number of incentives, among them ADS status

(Beldi, 2018). (2) Canada was the last Western country to receive ADS status (in 2009) because of Prime Minister Harper's criticisms of China' human rights accord and meeting with the Dalai Lama in 2007 (Tse, 2013). (3) The Republic of Palau never officially received ADS. This was not an issue between 2008 and 2017 where Chinese incoming tourist numbers surged from 634 to 87,000. However, by the end of 2017, a memo from Chinese officials was allegedly circulated to travel agencies, reminding them that Palau was not on the ADS list and that agencies would be penalised if they arranged travel to the country. This resulted in Chinese tourist numbers plummeting by 23% between the third and fourth quarter of 2017 – a tremendous hit for a country where 50% of GDP comes from tourism and 50% of incoming tourism comes from China. Observers reported that this was done to pressure Palau, who has maintained diplomatic ties with Taiwan rather than Mainland China (Beldi, 2018).

Finally, this soft power is also exerted indirectly, as many countries have incorporated growth in Chinese tourism numbers in economic development plans (Xu et al., 2020) and may avoid confrontation out of fear of repercussions.

The Chinese Tourism Law – How it propelled existing trends and enhanced the use of travel contracts

The swift growth of China's outbound tourism resulted in a number of challenges (Mak et al., 2011; Wang et al., 2018), which in turn prompted further tourism legislation from the Chinese government. Such a challenge were the so-called zero-fare or zero-commission tours (Chen et al., 2011; Zhang et al., 2009). Such tours are sold at a price corresponding to or even lower than the cost of making the trip. Profit is made through commission by "luring, cheating, or even forcing tourists into paying extra expenses and fees or in most cases through shopping (King et al., 2006; Zhang et al., 2009)" (Jørgensen et al., 2018, p. 4). To address the issues mentioned, the act was once again revised in 2009, where it was also renamed to 'Regulation on Travel Agencies'.

Specific rules included that travel agencies:

> Should not solicit tourists through a quoted price below the cost level; [...] not provide any other services than those specified in the contract; [...] not request tour guides to receive package tour groups without a due fee payment or with a lower-than-cost payment.
>
> (Huang, 2010, p. 158)

Notwithstanding these efforts, it seems that the malicious practices have persisted after 2009. Chen et al. (2011), for example, highlighted episodes of forced shopping reported in the media throughout 2010, which spurred continuous public debate. This may have been one of the main reasons behind the introduction of the 'Tourism Law of the People's Republic of China' also known as the 'Chinese Tourism Law'.

The 'Chinese Tourism Law' was effectuated on October 1, 2013 (CNTA, 2013a). It specified the rights of tourists, safety standards, rules for tour operators as well as guidelines for the handling of complaints and was hailed as a milestone because it allowed tourists to complain and seek redress for the first time (Jørgensen et al., 2020; South China Morning Post, 2013). It is notable how directly the law targeted specific malicious practices. An example of this is Article 41, which reads:

> Tour guides shall strictly follow the itinerary, shall not alter it or stop the services they are providing without permission or to ask for tips from tourists or lure, cheat, compel or force tourists in a disguised form to make purchases or participate in tourism activities that require additional payment.
>
> (CNTA, 2013, Article 41a)

Less than a handful of studies exist that deal specifically with the Chinese tourism law, meaning that information about its effectiveness is still lacking (Jinliang & Hong, 2013; Jørgensen et al., 2020; Ma et al., 2015; Wang et al., 2018). However, we will argue that the tourism law may have helped to propel existing behavioural trends among Chinese tourists – towards smaller groups, more focus on quality, self-driving and less focus on shopping. The main reason for this is that a result of the implementation of the 'Chinese tourism law' was an increase in tour prices (Global Blue Briefings, 2013; Schaal, 2013), which meant that the price gap between a normal (cheap) tour and a more in-depth trip was narrowed (Wang et al., 2018, p. 317).

Despite this positive outcome, there are also signs of problems with the law and how it is implemented. In 2019, Judge Tu Lin of the District court of Shunyi in Beijing commented to Xinhua news that the rights of travellers were being eroded by travel contracts. The judge urged travel consumers to carefully study and examine the content of the travel contract signed with travel agencies (Song & Wei, 2019). This could suggest limitations within the tourism law. The worry of the court could suggest that the travel agencies are protecting themselves against liabilities through travel contracts. The travel contract signed between the traveller and the travel agencies would in fact guide the relationship more than the tourism law itself because Chinese consumer law is very contract content-specific (Ge, 2019). This is a sign that China's outbound tourism is increasingly regulated through contracts between travel agencies and consumers rather than the (tourism) law, potentially eroding the legislation.

The past and future development of China's outbound tourism

The introduction of the ADS scheme and the consequent inclusion of more and more destinations combined with a relaxation of travel restrictions have produced a huge and ever-expanding flow of outbound Chinese tourists to countries all over the world (Cai et al., 2008). A growing economy and the increasing number of available holidays have also helped this development. Since the 1980s, China's economy has undergone explosive growth and is now among the largest in the world (International Monetary Fund [IMF], 2014). According to the National Bureau of Statistics China, per capita disposable income in China grew by an average of around 10% from 2012 to 2015. From this, a growing middle-upper class with increasing disposable income has emerged, for whom leisure travel is a possibility (Wang & Sheldon, 1995, as cited in Cai et al., 2001). The implementation of the five-day working week in the mid-nineties (Cai et al., 2008; Zhang & Qu, 1996, as cited in Kau & Lim, 2005), the 11 national holidays in the mid-noughties as well as the 5 to 15 days of paid annual leave that employees are now entitled to (de Sausmarez et al., 2012) also allowed many Chinese citizens time for travel and leisure.

Together, these factors make China one of the most rapidly expanding outbound tourism markets in the world. The most impressive growth rates of around 20% happened between 1991 and 2013 – from 6 million outbound tourists in 1994 to 98.19 million in 2013 (Chiang, 2012; CNTA, 2013b; ETC & UNWTO, 2013). Since then, the pace of growth has slowed slightly, but still presented an impressive growth to a total of 169.21 million in 2019. Based on this, it is still expected that China's outbound tourism will continue to grow at a healthy pace and from a larger base going forward.

Before the COVID-19 virus halted international travel almost completely, there were no signs that the growth of the China outbound market would stop in the near future. We find six important reasons for this. First, it is expected that the aforementioned economic growth and expansion of the middle-upper class will continue for years to come (Lin et al., 2015). Second, although the total number of Chinese people travelling abroad is high, it is still very low relative to the population – as an example, only around a tenth of the Chinese population owned a passport in 2019 (Huaxia, 2019). Third, the relatively low number of outbound tourists (162 million in 2019) compared to the high number of domestic tourists (6 billion in 2019) indicates potential for growth of the outbound tourism industry in the future (Statista, 2020a, 2020b). Fourth, over the next five to ten years, the largest cohort of Chinese white colour workers ever will retire. It is estimated that with the current retirement age of 60 for men and 55 for women somewhere between 250 and 300 million Chinese will retire in the period 2022 and 2030 (Feng et al., 2019). These retirees are the first generation with a strong purchasing power due to already acquired housing and good retirement benefits. Demographically, this suggests that the golden age of Chinese outbound tourism should begin in the 2020s. Fifth, outbound tourism is encouraged by the Chinese government because it can be used as a way to increase imports without reducing exports (Tse & Hobson, 2008) and can be used as a soft power tool, as previously explained. Sixth, the attractiveness of the Chinese market is rising with its continued maturation. Lim et al. (2020) explain that whereas the division between group travellers and (semi-)independent travellers was 50–50 in 2012. It is now around 20% group travellers, 40% independent travellers and 40% semi-independent travellers (where some parts of the journey are booked through agents).

Related to this, the Chinese outbound tourism market is also attractive from a macro-economic point of view, as it is key in terms of balancing China's trade balance with the rest of the world. This became evident in November 2020, when China presented its biggest trade surplus in decades, which was directly related to China's missing imports of tourism services due to the COVID-19 pandemic.

There is no doubt that the consequences of the COVID-19 pandemic on China outbound travel will be felt for years to come as airlines, hotels and travel agencies will either go out of business or consolidate. It is impossible to predict when the current environment will change, but the arguments above indicate a future of strong and continued growth for China's outbound tourism, albeit with a slight delay.

Conclusion

This chapter has explained how the 'Travel Agency Act' halted international competition and paved the way for OTAs such as Ctrip (now Trip.com); how the ADS scheme gave receiving countries channels to handle larger tourist groups, but also allowed outbound tourism to function as a soft power tool and how the 'Chinese Tourism Law' propelled existing travel trends and created a more rule-based arbitration system for travellers to refer to, as they became a significant consumer group, but also pushed agencies towards the use of travel contracts. Through these examples, this chapter has illustrated that while other tourism markets are primarily market-driven, China's outbound tourism has been and will continue to be affected by policy and legislation. From this, we conclude that in order to understand China's outbound tourism, one needs to understand its (tourism) policymaking.

Looking to the future, significant foreign investments in the Chinese tourism industry will likely not happen until the current investment restrictions are lifted or altered. Highly regulated sectors tend to strengthen already existing entities (OECD, 2018) and push out small and medium-size entities. Based on this knowledge, we foresee that the Chinese travel industry will likely continue to have few large entities such as the big OTAs like Trip.com and Tuniu that will dominate certain segments. These entities will serve as digital point of sales for smaller entities. This gives regulators the options of doing a trade-off between growth and stability. By having a few large operators, regulators can intervene to stabilize the market by steering the activities of these operators. This relationship will be at the core of the Chinese tourism industry's interaction with its government entities in the coming years of growth.

COVID-19 has stopped China's outbound tourism in its tracks, and the big questions are if, how and when it will bounce back. We have presented six reasons that it will come back strong and call for further research into precisely how and when.

References

Arita, S., La Croix, S., & Mak, J. (2012). How China's approved destination status policy spurs and hinders Chinese travel abroad. *UHERO Working Paper, No. 2012–6R*. https://www.asianstudies.org/publications/eaa/archives/how-chinas-approved-destination-status-policy-spurs-and-hinders-chinese-travel-abroad/

Beldi, L. (2018, August 25). China's 'tourist ban' leaves Palau struggling to fill hotels and an airline in limbo. *ABC News*. https://www.abc.net.au/news/2018-08-26/china-tourist-ban-leaves-palau-tourism-in-peril/10160020

Cai, L. A., Lehto, X. Y., & O'leary, J. (2001). Profiling the U.S.-bound Chinese travelers by Purpose of trip. *Journal of Hospitality & Leisure Marketing*, *7*(4), 3–16. doi:10.1300/J150v07n04_02

Cai, L. A., Li, M., & Knutson, B. J. (2008). Research on China outbound market: A meta-review. *Journal of Hospitality & Leisure Marketing*, *16*(1–2), 5–20. doi:10.1080/10507050802096778

Chen, Y., Mak, B., & Guo, Y. (2011). "Zero-fare" group tours in China: An analytic framework. *Journal of China Tourism Research*, *7*(4), 425–444. doi:10.1080/19388160.2011.627024

Chiang, M.-H. (2012). The changing role of tourism in China's economy. *Journal of China Tourism Research*, *8*(2), 207–223. doi:10.1080/19388160.2012.677372

China National Tourism Administration. (2013a). *Tourism law of the People's Republic of China*. Beijing: China National Tourism Administration.

China National Tourism Administration. (2013b). *China National Tourism Administration Statistics 2013*. http://cnta.gov.cn/

Ctrip. (2003, November 13). *Investor relations. SEC filling. Registration statement for certain foreign private issuers*. http://ir.ctrip.com/phoenix.zhtml?c=148903&p=irol-sec

Dai, B., Jiang, Y., Yang, L., & Ma, Y. (2017). China's outbound tourism – Stages, policies and choices. *Tourism Management*, *58*, 253–258. doi:10.1016/j.tourman.2016.03.009

Dai, X. F. (2011). Outbound tourism should become an important mean to expand China's international influence. *Tourism Tribune*, *26*(8), 6–7.

de Sausmarez, N., Tao, H., & McGrath, P. (2012). Chinese outbound tourism to the United Kingdom: Issues for Chinese and British tour operators. *Journal of China Tourism Research*, *8*(3), 268–283. doi:10.1080/19388160.2012.704235

European Travel Commission, & United Nations World Tourism Organization. (2013). *The Chinese outbound travel market-2012 update*. https://www.e-unwto.org/doi/book/10.18111/9789284414307

Feng, Q., Yeung, W.-J. J., Wang, Z., & Zeng, Y. (2019). Age of retirement and human capital in an aging China, 2015–2050. *European Journal of Population*, *35*(1), 29–62. doi:10.1007/s10680-018-9467-3

Ge, J. (2019). *A comparative analysis of policing consumer contracts in China and the EU.* Singapore: Springer Singapore.

Gillis, P. (2011, March 20). *Explaining VIE structures.* http://chinaaccountingblog.com/weblog/explaining-vie-structures.html

Global Blue Briefings. (2013). *Global blue briefings: Special edition on the new Chinese National Tourism Law.* Vaud: Global Blue.

Guo, Y., Seongseop Kim, S., & Timothy, D. J. (2007). Development characteristics and implications of Mainland Chinese outbound tourism. *Asia Pacific Journal of Tourism Research, 12*(4), 313–332. doi:10.1080/10941660701760995

Huang, S. (2010). Toward a behavioral theory of government firm relationship behavior: Thick description of the dynamics of government's role in shaping China's domestic, inbound, and outbound tourism industry. In A. G. Woodside (Ed.), *Tourism-marketing performance metrics and usefulness auditing of destination websites* (pp. 149–163). Bingley: Emerald Group Publishing.

Huaxia. (2019, September 28). China Focus: 70 years on, Chinese travel abroad more easily in much larger number. *Xinhuanet.* http://www.xinhuanet.com/english/2019-09/28/c_138430646.htm#:~:text=The%20number%20of%20Chinese%20passport, to%2030.08%20million%20in%202018

International Monetary Fund. (2014). *World economic outlook database.* http://www.imf.org/external/pubs/ft/weo/2011/02/weodata/index.aspx

Jinliang, C., & Hong, Y. (2013). A comprehensive exploration of the influence Tourism Law has on Tourists' Impolite behavior (Translated from Chinese). *Journal of Exploration, 12*, 219–220.

Jørgensen, M. T., King, B. E., & Law, R. (2020). The impacts of China's policy-making and legislation on outbound tourism – perspectives from long-haul intermediaries. *Journal of Policy Research in Tourism, Leisure and Events, 12*(1), 21–33. doi:10.1080/19407963.2018.1505104

Kau, A. K., & Lim, P. S. (2005). Clustering of Chinese tourists to Singapore: An analysis of their motivations, values and satisfaction. *International Journal of Tourism Research, 7*(4), 231–248. doi:10.1002/jtr.537

King, B., Dwyer, L., & Prideaux, B. (2006). An evaluation of unethical business practices in Australia's China inbound tourism market. *International Journal of Tourism Research, 8*(2), 127–142. doi:10.1002/jtr.563

Kristensen, A. E. (2020). Foreign direct investment in the Chinese outbound tourism industry: The policies, participants and possibilities. *Journal of Policy Research in Tourism, Leisure and Events, 12*(1), 8–20. doi:10.1080/19407963.2018.1505100

Lim, D. J., Ferguson, V. A., & Bishop, R. (2020). Chinese outbound tourism as an instrument of economic statecraft. *Journal of Contemporary China, 29*, 1–18. doi:10.1080/10670564.2020.1744390

Lin, S., Liu, A., & Song, H. (2015). Modeling and forecasting China outbound tourism: An econometric approach. *Journal of Travel and Tourism Marketing, 32*(1–2), 34–49. doi:10.1080/10548408.2014.986011

Ma, E., Qu, C., Hsiao, A., & Jin, X. (2015). Impacts of China Tourism Law on Chinese outbound travelers and stakeholders: An exploratory discussion. *Journal of China Tourism Research, 11*(3), 229–237. doi:10.1080/19388160.2015.1077182

Ma, J. X., Buhalis, D., & Song, H. (2003). ICTs and Internet adoption in China's tourism industry. *International Journal of Information Management, 23*(6), 451–467. doi:10.1016/j.ijinfomgt.2003.09.002

Mak, A. H. N., Wong, K. K. F., & Chang, R. C. Y. (2011). Critical issues affecting the service quality and professionalism of the tour guides in Hong Kong and Macau. *Tourism Management, 32*(6), 1442–1452. doi:10.1016/j.tourman.2011.01.003

Organisation for Economic Co-operation and Development. (2018). *Improving the business environment for SMEs through effective regulation* (SME Ministerial Conference) [Policy Note]. https://www.oecd.org/cfe/smes/ministerial/documents/2018-SME-Ministerial-Conference-Parallel-Session-1.pdf

Pan, G. W., & Laws, E. (2003). Tourism development of Australia as a sustained preferred destination for Chinese tourists. *Asia Pacific Journal of Tourism Research, 8*(1), 37–47. doi:10.1080/10941660308725454

Qu, H., & Lam, S. (1997). A travel demand model for Mainland Chinese tourists to Hong Kong. *Tourism Management, 18*(8), 593–597. doi:10.1016/S0261-5177(97)00084-8

Schaal, D. (2013, November 27). The independent Chinese tourist gets a boost from clampdown on packaged tour abuses. *Skift*. http://skift.com/2013/11/27/china-tourism-law-hurts-packaged-tours-boosts-independent-chinese-travelers/

Song, X., & Wei, X. (2019, August 30). In the first half of this year, a Beijing court heard more than 484 cases regarding travel disputes. Judge reveals "The pitfalls when booking group tours". *Xinhua*. http://www.xinhuanet.com/local/2019-08/30/c_1124939226.htm

South China Morning Post. (2013, May 13). China's tourism law will benefit everyone. *South China Morning Post*. http://www.scmp.com/comment/insight-opinion/article/1236226/chinas-tourism-law-will-benefit-everyone

Statista. (2020a). *China: Number of domestic tourist arrivals 2019*. Statista. https://www-statista-com.ep.fjernadgang.kb.dk/statistics/277254/number-of-domestic-trips-in-china/

Statista. (2020b). *China: Outbound journeys of Chinese tourists 2019*. Statista. https://www-statista-com.ep.fjernadgang.kb.dk/statistics/277250/number-of-outbound-journeys-of-chinese-tourists/

Tsang, N. K. F., & Hsu, C. H. C. (2011). Thirty years of research on tourism and hospitality management in China: A review and analysis of journal publications. *International Journal of Hospitality Management, 30*(4), 886–896. doi:10.1016/j.ijhm.2011.01.009

Tse, T. S. M. (2011). China's outbound tourism as a way of ordering. *Journal of China Tourism Research, 7*(4), 490–505. doi:10.1080/19388160.2011.627031

Tse, T. S. M. (2013). Chinese outbound tourism as a form of diplomacy. *Tourism Planning & Development, 10*(2), 149–158. doi:10.1080/21568316.2013.783738

Tse, T. S. M., & Hobson, J. S. P. (2008). The forces shaping China's outbound tourism. *Journal of China Tourism Research, 4*(2), 136–155. doi:10.1080/19388160802279459

Wang, W.-F., Chang, Y., & Pearce, P. L. (2018). China's first Tourism Law: Representations of stakeholders' responses. *Journal of Tourism and Cultural Change, 16*(3), 309–327. doi:10.1080/14766825.2017.1298607

Xu, H., Wang, K., & Song, Y. M. (2020). Chinese outbound tourism and soft power. *Journal of Policy Research in Tourism, Leisure and Events, 12*(1), 34–49. doi:10.1080/19407963.2018.1505105

Zhang, H. Q. (2004). Accession to the World Trade Organization: Challenges for China's travel service industry. *International Journal of Contemporary Hospitality Management, 16*(6), 369–372. doi:10.1108/09596110410550815

Zhang, H. Q., Chong, K., & Ap, J. (1999). An analysis of tourism policy development in modern China. *Tourism Management*, *20*(4), 471–485. doi:10.1016/S0261–5177(99)00020-5

Zhang, H. Q., Heung, V. C. S., & Yan, Y. Q. (2009). Play or not to play—An analysis of the mechanism of the zero-commission Chinese outbound tours through a game theory approach. *Tourism Management*, *30*(3), 366–371. doi:10.1016/j.tourman.2008.08.006

3 From then to now

A brief history of China's outbound tourism market

Vorakarn Chalermchaikit and Yixuan Cheng

Introduction

During the recent years, global perception has witnessed the rapid development of China's outbound tourism. This development has attracted a great deal of attention from tourism researchers. Understanding China's outbound tourism and tourists has therefore become an increasingly important issue. Although the Chinese citizens have a relatively short history of traveling abroad, the number of Chinese outbound tourists has increased rapidly, enabling it to be a significant market in the world. Due to the growing potential of China's outbound tourism, numerous studies have been conducted to explore Chinese tourists' motivation and behaviors. However, the previous research has tended to consider the Chinese outbound tourism market as being homogenous (Dai et al., 2016). Such assumption ignored the differences in the generations of the tourists and their respective behavioral characteristics.

It can be inferred that it necessitates a deeper recognition that Chinese outbound tourists constitute a set of subgroups rather than a single mass. Generational analysis has been perceived as a useful segmentation method because the relatively homogeneous characteristics exist within generations, and obvious differences are demonstrated across generations (Schewe & Noble, 2000). Despite that the generational analysis has been widely used in tourism studies, seldom research has been conducted to investigate the Chinese outbound market from a generational perspective. Moreover, the outbreak of COVID-19 has brought an adverse consequence on the global tourism industry and also affected Chinese tourists' consumption patterns in terms of outbound tourism. Therefore, a critical analysis is urgently needed to predict the future trends of Chinese outbound tourism under the influence of COVID-19.

Thus, to address the identified gaps existing in the literature and provide a thorough understanding of the Chinese outbound tourism market, the objective of this chapter is to comprehensively analyze the behavioral characteristics of the Chinese outbound tourists. To achieve this research objective, this chapter interprets chronologically with rational information about

DOI: 10.4324/9781003121329-4

the development of Chinese tourism with the aid of rational information. The first section introduces the history of Chinese outbound tourism. Then, a specific comparison of Chinese outbound tourists' motivation between elder generation and young generation is demonstrated in the second section. Finally, based on the understanding of Chinese outbound tourism history and tourists' motivation, a prediction of future trend of Chinese outbound tourism is summarized in the third section.

The history of Chinese outbound tourism

The notion of travel is traditionally deep-rooted in Chinese culture, which is also reflected in Chinese literature. Traveling to see the outside world is perceived as an important part of life among Chinese people (Tse & Hobson, 2008). China has currently become a major of global tourism industry because the national economy has grown rapidly. The image of Chinese tourists can be distinguished globally and widely. The growing phenomenon of Chinese outbound tourism has been attractive through the global tourism industry.

According to Johnson et al. (2020), the development of China's outbound tourism can be divided into four stages. The first stage (1983–1997) refers to 'family visits'. In 1983, Mainland Chinese citizens were allowed to visit the Special Administration Regions of Hong Kong and Macao as well as selected Southeast Asian countries to see friends and relatives. This marked the beginning of Mainland Chinese outbound tourism (Johnson et al., 2020). During the second stage (1997–2005), Chinese tourists could travel abroad to approved destinations with the implementation of Approved Destination Status (ADS) agreements (Johnson et al., 2020).

The third stage (2005–2011) witnessed the maturing of China's outbound tourism, where Chinese tourists were not restricted with limited choices of destinations; instead, they were provided with a wide range of selections (Johnson et al., 2020). The fourth stage (2012–present) is characterized by great Chinese government support for outbound tourism development. An increasing number of destination countries reduced visa restrictions to attract Chinese tourists (Johnson et al., 2020).

Previous studies claimed that Chinese outbound tourism has been developing the number of tourists and generating the amount of expenditure which are perceived as a major phenomenon of the global tourism (Law et al., 2016). Furthermore, other previous academic studies mentioned that the fast growth of China's outbound tourism market is held up by China's economic reform and open-door policy with relaxation of visa restrictions stated by Chinese government along with the growth of Chinese' incomes (Huang, Keating et al., 2015; Jørgensen et al., 2017).

Chinese citizens have a very short history of traveling abroad. In ancient period of China, only representatives and diplomats of the country have had the opportunity to travel abroad, with purposes of cultural communication and national image dissemination (Xu et al., 2020). In the first three decades of the People's Republic of China, outbound tourism was limited to a few officers. With economic growth and political stability, the Chinese government changed the outbound tourism policy from being limited to allow the citizens to travel abroad (Xu et al., 2020). China is now the world leader in tourism market, and Chinese tourists are placed at the top of the list globally (United Nations World Tourism Organization, 2014).

The growth of Chinese outbound tourism

The primary of economic growth of China was first implemented after the open-door policy in 1978, and also the Chinese government relaxation's policy was introduced via visa restrictions with the increasing wealth of Chinese population (Huang, Li et al., 2015; Zhong et al., 2015). Before this policy was launched, Chinese could travel overseas for official reasons only such as attending athletic meets and diplomatic events (Arlt, 2013).

However, since 1983 onwards, after the implementation had been launched effectively, Chinese tourists had been allowed to travel aboard for leisure purposes (Law et al., 2016). This phenomenon was a positive sign encouraging Chinese to travel abroad. In 1997, the Chinese government signed the first bilateral agreements with non-Asian countries including Australia and New Zealand, providing Chinese citizens another opportunity to travel and experience Western countries (Keating & Kriz, 2008). Furthermore, another highlighted incidence occurred in 1999 when Chinese were permitted to travel to non-Asian countries, where all the 117 countries had been approved (Law et al., 2016).

Later on, in 2003, Individual Visitor Scheme (IVS) had been launched to be applicable for Chinese, and at the end of 1990s, the bilateral agreements between three Southeast Asian countries, Malaysia, Singapore and Thailand, and China were applied to implement an ADS (Law et al., 2016). Chinese outbound tourism had then been managed and regulated by ADS system. This system concerns bilateral agreements between China and selected overseas destinations. Hence, the ADS system restricts the overseas destinations to where Chinese can travel and which international tourism companies are allowed to operate in China (Keating & Kriz, 2008).

In 2005, China was characterized as the country impacting global travel experience, and also in 2011, the second wave of China's outbound tourism has begun (Huang & Xu, 2018). During this period, Chinese who are elites middle-age tend to support the outbound travel because they have high demand to experience traveling outside and purchasing power. Therefore, they can support themselves and their families, which are considered as the great opportunities for European and other countries to welcome Chinese tourists.

Due to this rapid growth of Chinese outbound tourism, it also provides Chinese international students with a wide range of travel opportunities aboard. In general, Chinese international students stay for longer period than mainstream Chinese tourists. Therefore, they have more opportunities to travel to many destinations where they can visit either in that country and others. In addition, some studies state that the major contribution via tourism industry of the host country starts when Chinese international students spend their holidays during their semester break and after the completion of their major studies (Gardiner, King, & Wilkins, 2013; Huang & Tian, 2013).

In summary, multiple macro-environment aspects including economic, cultural, political and technological apparently shape the outbound tourism development in China (Dai et al., 2017; Guo et al., 2007; Tse & Hobson, 2008). Moreover, after the implementation of "open-door" policy in 1978, China has been embarking on a series of reforms and liberalization. The state-owned travel agencies and companies were reformed to become consumer-oriented and market-driven (Tse & Hobson, 2008). The tourism industry has become very active with the growth of demands in markets. Many destinations aimed for Chinese travelers and they try to promote their destinations in China (Guo et al., 2007; Tse & Hobson, 2008).

In addition, the technological information's advancement makes destination information be easily available to potential Chinese tourists. In an extensive country like China, it seems that internet is an effective approach to promote travel products. There are already many websites and social media offering travel information and services. All of these factors encourage the outbound international tourist flows from China positively (Dai et al., 2017; Tse & Hobson, 2008). Also, with the high rate of disposable income and leisure time, it can be inferred that more Chinese people will aim to travel abroad.

The comparison of Chinese outbound tourists' motivation between generation X and generation Y

The theory of generation was originally developed by Mannheim (Mannheim, 1952). A generation means a group of people who were born in the same period and experienced the same significant lift events. Thus, they acquire collective memory, and at the same time, they develop similar values and behaviors in a lifetime. To illustrate, generation theory has been widely applied in different social areas, including tourism studies (Huang & Lu, 2017). It shows that generational analysis can be applied effectively as a valuable segmentation technique for tourism research, which helps to deeply understand the backgrounds of tourist motivations (Huang & Lu, 2017). According to the differences among generations, tourism marketers can scope the market segment, recognize the needs of targeted market, design products and services, and create appropriate marketing strategies (Huang & Lu, 2017).

The new generation has been known as Millennials describing the new concept throughout the thinking and reaction (Crişan, 2016). Moreover, the academic study claims that the segment of individuals for people who were born during 1980 to 2000 are considered as generation Y and is characterized by the type of personalities as flexible, liberal, enthusiastic and impatient behaviors because they grow up under the era of technological development (Weingarten, 2009). Consequently, they have high expectations to gain good salaries in order to spend them on products and services with the strong willingness to get updates from new technologies (Jain & Pant, 2012). It can be inferred that they may have more opportunities to show their lifestyle when they travel because social media becomes an important part of their lives.

Also, they expect to get everything as fast as possible and prefer to try new things that they perceive directly their personality and preference. The mainstream type of travel may not be perceived as an attractive travel if they feel uninterested in those activities. However, if they feel happy and satisfied, they tend to spread the positive words of mouth from their traveling activities via social media such as Facebook, Twitter and Instagram. This would reflect the different perceptions via tourists from other elder generations.

For instance, generation X is the generation of people who were born between 1965 and 1980 (Weingarten, 2009). This generation has encountered by many political situations globally such as the Vietnam War and the Cold War. Hence, they tend to be open to the diversity and value while focusing on their work life balance between family and work. Furthermore, another generation, the Baby Boomer, represents people who were born after the World War II between 1946 and 1964 (Weingarten, 2009). They have strong value in traditional working style referring to the strict patterns and high endeavors (Wright, 2005).

Generation defines the group of people who share their proximities in age, experiences in history or development stages (Benckendorff et al., 2010). Also, the theory of generation states that not only age but also the years of individuals have impact in grooming behavioral patterns (Gardiner, King, & Grace, 2013.) Hence, travel patterns can be varied from behaviors and attitudes based on the different generation of Chinese tourists who were raised in different generations as China has many specific historical incidences.

Therefore, it can affect behavioral pattern differing from general generation's pattern. Generation X in China are individuals who were born during the huge propaganda era of Cultural Revolution (Yang & Lau, 2015). Currently, Chinese who are Gen X are in their mid-30s to 50s. They have many memories of political instability with numerous social values. They are excited about the national development and the influx of new information with both economic and academic opportunities (Yang & Lau, 2015). In addition, they are a hard-working and pragmatic generation. They are accustomed to saving their money rather than spending it, which affects their traveling style. They prefer to make sure everything and minimize uncertainty avoidance, so they tend to choose travel packages that they can travel in groups and all facilities are provided already.

The changes of Chinese outbound travel pattern among different generations

On the other hand, Generation Y in China, also known as the post-1980s, are born after the Chinese government legislating the One-Child Policy. They hold different values from Gen X-ers because they are raised well and provided with better conditions. Most of them are independent travelers who prefer control on their travel plan and pace (So et al., 2016). Meanwhile, these independent young travelers have been characterized as experienced tourists, well-educated with good language skills and self-challengers (So et al., 2016). Prayag et al. (2015) claim that these young travelers have different service expectations comparing with old generations in many ways such as a preference for dinning outside of hotels and experiencing local authentic cultures.

Apart from the high intelligence on an academic level, Gen Y-ers in China might be perceived as demanding, ambitious and intolerant practices. However, they are characterized as optimists toward their future, new excitement for consumerism and acceptance of the role via China's transformation from modern to superpower (Jin et al., 2014). From these different generations, it can be seen that both generations show the different traits via their unique experiences when they travel. The 'New Chinese Tourist' is categorized by Gen Y-ers who prefer to engage in in-depth travels in exotic destinations (Arlt, 2013). Also, they are active in sharing their travel experience and seeking updates on online platforms and social media where they rely on.

For baby boomers, with only One-Child Policy in the family, they do not have strong preference and demand to save their money for future generations. Therefore, they are inclined to take their leisure from their travel after their retirement. As they have money and time, they tend to travel to many destinations where they spend times with their families and friends. At the same time, their children may want to show their respect and take care for their parents, so they tend to support their parent's travel demands.

The travel pattern shows the distinguishment between different generations. Chinese senior tourists have more free time than younger generations, and they prefer to travel during low seasons (Nasolomampionona, 2014). However, the biggest challenge they faced is lack of language skills (Kairos Future, 2016). Meanwhile, due to lack of international travel experience, most of them rely on travel agencies and friends strongly to gain outbound travel information via some applications such as Ctrip. Therefore, they prefer to select organized group tours to travel abroad or are accompanied by their family who can speak English. As they grew up during the end of World War II period, they seem to be attracted by the history.

Consequently, they prefer experiencing authenticity while they travel to touch the local feeling and absorb the history at the same time rather than shopping. Even though their travel expenditure may not be as high as the young generation tourists preferring shopping and following the new trends, they value highly in cultural experiences and local tradition. Moreover, they also focus on safety issues (Bao, 2009) and may avoid the big events. Because the new trend of Chinese senior tourists become more active in the leisure travel (Li et al., 2014), it influences on the massive market.

The future trend of Chinese outbound tourism

Based on the analysis of Chinese outbound tourists' motivations and behavior characteristics, some suggestions were given to the tourism marketers and tour operators (Huang & Lu, 2017). First, cultural elements especially the authentic ones should be highlighted in the promotions of destinations because it is one of the main factors motivating Chinese people to travel abroad (Huang & Lu, 2017). Second, raised standards and high-quality services should be provided to attract Chinese tourists. Third, safety assurance should be highly considered, as Chinese tourists have increasingly taken safety into consideration when they travel outbound (Huang & Lu, 2017). In terms of information distribution, electronic word-of-mouth on websites and social media is recommended to distribute information for Chinese market. Because Chinese tourists prefer to acquire travel information on internet (Huang & Lu, 2017)

Additionally, from the cultural perspective, Chinese people give the importance through collectivism which is a unique feature of the country (Fan, 2000). Therefore, based on their personal level, they regard themselves as a part of a group, so they have high concerns about in-group objectives which they help to generate benefits and interests. A collectivist orientation is conveyed firmly by obligations and norms with group collaboration although the individual benefits are not clear (Ravlin et al., 2012).

Therefore, it implies that individuals are motivated by their desire and preference to be similar to other members and influence on collective similarity where they are expected to prioritize and commit the group's goals more than their personal objectives. This collective principle plays an important role in Chinese individual's lifestyle. For instance, traditionally, Chinese tourists prefer all-inclusive travel packages where they can travel in groups to visit famous destinations. Also, they tend to travel during their public holidays to visit their families such as the Chinese New Year and the Spring Festival.

Furthermore, they prefer to spend their time with their parents or friends during events and festivals or intimate atmosphere. They enjoy to dine with other people to connect for personal or for special occasions (Ibrahim & Howe, 2011). Generally, Chinese tend to order many dishes to share together rather than order individually because during the dining time, it is a special time to connect and feel each other while they eat, drink and talk. They use the same utensils to serve themselves. Also, they serve food in small portions to offer hospitality and building relationships.

Therefore, from this traditional practice, it reflects and motivates the tourist motivations and characteristics to have some partial activities in-groups. Even though different generations have different behaviors, they are still linked by their collectivist culture. The elder tourists enjoy traveling in group because they do not want to encounter with uncertain situations that may affect their traveling experiences due to the language barrier and inexperience. Hence, they tend to go out and stick in groups to secure themselves and feel taken care of by others. However, for younger tourists, even though most of them do not have language barriers as the elder generation, they express collective orientation via social media. Although they travel in small groups or with partners, they still search and look at social media for reviews and opinions from other Chinese tourists who already have experiences in those destinations.

The practical trend of Chinese outbound tourism during pandemic era

In 2020, the global pandemic COVID-19 has made an enormous impact on various businesses. Due to the pandemic situation associated with COVID-19, the travel industry has direct impact and consequences. The number of tourists on global level has declined. It is estimated by the World Tourism and Travel Council (WTTC) that COVID-19 may have caused a huge loss on the global tourism industry, amounting to a minimum of 22 billion dollars (Zhu & Deng, 2020).

Meanwhile, COVID-19 may have far-reaching influences on tourists' preferences and consumption patterns. People tend to travel in domestically rather than aboard. Therefore, Chinese tourists seem to travel independently or in smaller groups and take only short trips because they would like to remain safe and secure. They try to avoid crowded destinations, and they may prefer to visit less well-known destinations in China.

Moreover, they try not to travel during the Chinese public holidays as before. Thus, the tourism pattern has changed extremely, and the destinations have to adjust themselves too. However, this is a good opportunity for local destinations to promote their cultures and authenticities as 'unseen China' because they can capture awareness of Chinese tourists to visit unique places where not many people have visited before. This could change the pattern to minimize the over tourism or the massive travel group and promote the sustainability in long term.

According to Zhu and Deng (2020), with safety measures taken into account in the context of COVID-19 pandemic, Chinese tourists tend to travel in rural destinations for relaxation. Rural tourism shows a series of advantages, such as low cost, low flow density and short time consumption, which fulfills the psychological needs of tourists to enjoy natural scenery and guarantees safety in the context of COVID-19 (Zhu & Deng, 2020). In addition, having experienced the longer shutdown period of travel industries, people pay more attention to rural tourism for meeting their travel demands during this exceptional period.

Moreover, Zhu and Deng (2020) argue that perceived risk is more convincing in affecting travel intention rather than perceived quality in the epidemic period. Therefore, there is a suggestion that tourism destinations need to give careful considerations to tourists' risk perception and make an appropriate plan for the risk management (Zhu & Deng, 2020). Destinations can reduce the risk perceived by tourists through increasing the COVID-19 prevention-related knowledge of tourists. For example, tourism managers can use social media to spread information on epidemic prevention measures that tourists should pay attention to during their traveling (Zhu & Deng, 2020).

Conclusion

In summary, it is widely recognized that Chinese outbound tourism development has experienced a rapid increase. Since the national reform and open-door policy was implemented in 1978, China's economy has been rapidly developed, resulting in improvements of living standard for most Chinese people. With the increase in personal disposal income of Chinese citizens, the demand for outbound travel has been increased dramatically. Meanwhile, the government's lessening the control of Chinese citizens' traveling abroad is also a key factor driving the development of Chinese outbound tourism. Moreover, the advances in information technology make destination information and online booking being available for most tourists, promoting the development of outbound tourism in China.

From the market perspective, the rapid growth of China's outbound tourism market has attracted numerous attentions. China has emerged as an important export market for many countries. Given the importance of the Chinese outbound market, it is significant to have a better understanding of traveling patterns, holiday preferences and tourism expectations of Chinese tourists, thereby developing effective strategies to capture the market. Therefore, the study explores and analyzes the motivations and behavior of Chinese outbound tourists.

It shows that there exists difference in preferences and characteristics between generation X and generation Y of Chinese outbound tourists. Generation X are accustomed to being more cautious about spending money on travel. They prefer to minimize uncertainty and choose travel packages. Comparatively, generation Y are characterized as well-educated, experienced and self-challenged. Most of them are more independent and prefer to control their travel pace.

Additionally, it was found that young travelers show different expectations toward outbound tourism compared to the older generations in many ways. The previous studies tend to approach Chinese tourists as a single homogenous group. However, it argues that Chinese tourists do not form a single homogenous group. For example, generation X and generation Y show differences in terms of preference, expectations and behaviors. Therefore, it suggests that a distinction among Chinese outbound tourists should be made thereby generating appropriate strategies to the targeted market. With the rapid development in economy and increase of leisure time, it can be predicted that more Chinese people will travel abroad in the future.

However, under the influence of global pandemic COVID-19, the number of tourists shows a dramatic decline globally. Meanwhile, it demonstrates that COVID-19 may influence tourists' preference and consumption patterns. Therefore, it can be inferred that COVID-19 brings up some challenges toward the tourism industry directly since the year 2020, but it refers to good signs for rebuilding Chinese tourism market to look at the patterns of young generation.

Moreover, stakeholders in the travel industry may take this opportunity to revise the necessary elements in order to improve tourists' experiences throughout before, during and after they visit particular destinations. It is not only the host of each destination who has to adjust to the new changes, but the path of tourism is also changed by the market in the eyes of domestic tourists, which can be considered as lessons of experience improvements in the markets. This is a positive outcome for domestic travel because people can gain confidence effectively to travel within China, where they can have wide chances to visit far destinations where they have never been to before or may not know of. Hence, it enhances self-guiding or small group traveling that tourists tend to rely on themselves to look for information on social media rather than buying a package which they have to travel in a big group with tour guide.

Thus, this phenomenon can be considered as a new scenery of a new tourism era referring to a new type of travelers because this pandemic can grab awareness of Chinese tourists to focus on domestic scenery. They can start looking for other authentic destinations that are not crowded as in urban cities with high risk of COVID-19 spreading. They are more interested to go to natural spots because they pay more attention to their health during the trips.

Also, they can consider road trips rather than groups because they have more privacy and comfort when they travel with their families or close friends. Apart from road trips, some of them may choose to spend time with their families at home, so they can decide to travel to meet their families and devote more times while they are more aware of the importance of family relations. As safety has been perceived as an important factor by Chinese tourists, it has been more convincing in influencing their travel intentions. Rural tourism in this epidemic period shows a series of advantages, including low travel cost, low flow density and peaceful and natural scenery. Therefore, Chinese tourists tend to choose rural destinations for relaxation.

The study briefly introduces the development of Chinese outbound tourism and analyzes the characteristics of different generation of outbound tourists, which provides a comprehensive understanding of Chinese outbound tourists and also help tourism managers create effective strategies to the targeted market. Furthermore, the study also predicts the future trends of Chinese tourists' preference and behavior based on the changes in external factors such as the global pandemic. Such prediction further helps practitioners to adjust their tourism policies and strategies accordingly.

References

Arlt, W. G. (2013). The second wave of Chinese outbound tourism. *Tourism Planning & Development, 10*(2), 126–133. doi:10.1080/21568316.2013.800350

Bao, Y. F. (2009). A study on seniors' travel motivation in Hangzhou based on push-pull theory. *Tourism Tribute, 24*(11), 47–52.

Benckendorff, P. J., Moscardo, G., & Pendergast, D. (Eds.). (2010). *Tourism and generation Y.* London: CABI.

Crişan, L. C. (2016). Generation Y-The management conversion paradigm. *Review of International Comparative Management, 17*(1), 76–86. http://rmci.ase.ro/no-17vol1/08.pdf

Dai, B., Jiang, Y., Yang, L., & Ma, Y. (2017). China's outbound tourism – Stages, policies and choices. *Tourism Management, 58*, 253–258. doi:10.1016/j.tourman.2016.03.009

Fan, Y. (2000). A classification of Chinese culture. *Cross Cultural Management: An International Journal, 7*(2), 3–10. doi:10.1108/13527600010797057

Gardiner, S., King, C., & Grace, D. (2013). Travel decision making: An empirical examination of generational values, attitudes, and intentions. *Journal of Travel Research, 52*(3), 310–324. doi:10.1177/0047287512467699

Gardiner, S., King, B., & Wilkins, H. (2013). The travel behaviours of international students: Nationality-based constraints and opportunities. *Journal of Vacation Marketing, 19*(4), 287–289. doi:10.1177/1356766712471233

Guo, Y., Seongseop Kim, S., & Timothy, D. J. (2007). Development characteristics and implications of Mainland Chinese outbound tourism. *Asia Pacific Journal of Tourism Research, 12*(4), 313–332. doi:10.1080/10941660701760995

Huang, Q., & Lu, Y. (2017). Generational perspective on consumer behavior: China's potential outbound tourist market. *Tourism Management Perspectives, 24*, 7–15. doi:10.1016/j.tmp.2017.07.008

Huang, R., & Tian, X. (2013). An investigation of travel behaviour of Chinese international students in the UK. *Journal of China Tourism Research, 9*(3), 277–291. doi:10.1080/19388160.2013.812898

Huang, R., & Xu, T. (2018). Opportunities and challenges brought by the China outbound travel market. *Worldwide Hospitality and Tourism Themes, 10*(6), 642–651. doi:10.1108/whatt-07–2018–0049

Huang, S. (Sam), Keating, B. W., Kriz, A., & Heung, V. (2015). Chinese outbound tourism: An epilogue. *Journal of Travel & Tourism Marketing, 32*(1–2), 153–159. doi:10.1080/10548408.2014.986018

Huang, Z. (Joy), Li, M., & Li, Q. (2015). An examination of Taiwan destination brand associations: From the perspective of Mainland Chinese tourists. *Journal of Travel & Tourism Marketing, 32*(1–2), 50–64. doi:10.1080/10548408.2014.986012

Ibrahim, R. W., & Howe, D. (2011). The experience of Jordanian care leavers making the transition from residential care to adulthood: The influence of a patriarchal and collectivist culture. *Children and Youth Services Review, 33*(12), 2469–2474. doi:10.1016/j.childyouth.2011.08.019

Jain, V., & Pant, S. (2012). Navigating generation Y for effective mobile marketing in India: A conceptual framework. *International Journal of Mobile Marketing, 7*(3), 56–65.

Jin, T., Lin, V. S., & Hung, K. (2014). China's generation Y's expectation on outbound group package tour. *Asia Pacific Journal of Tourism Research, 19*(6), 617–644. doi:10.1080/10941665.2013.806939

Johnson, P. C., Xu, H., & Arlt, W. G. (2020). Outbound Chinese tourism: Looking back and looking forward. *Journal of Policy Research in Tourism, Leisure and Events, 12*(1), 1–7. doi:10.1080/19407963.2018.1505098

Jørgensen, M. T., Law, R., & King, B. E. (2017). Understanding the past, anticipating the future—A critical assessment of China outbound tourism research. *Journal of Travel & Tourism Marketing, 34*(7), 880–891. doi:10.1080/10548408.2016.1220889

Kairos Future. (2016, December). Tourism flows from China to the European Union–Current state and future developments. *European Commissions.* https://op.europa.eu/en/publication-detail/-/publication/bda9c8a2-4c43-11e7-a5ca-01aa75ed71a1/language-en/format-PDF

Keating, B., & Kriz, A. (2008). Outbound Tourism from China: Literature Review and Research Agenda. *Journal of Hospitality and Tourism Management, 15*(1), 32–41. doi:10.1375/jhtm.15.32

Law, R., Sun, S., Fong, D. K. C., Fong, L. H. N., & Fu, H. (2016). A systematic review of China's outbound tourism research. *International Journal of Contemporary Hospitality Management, 28*(12), 2654–2674. doi:10.1108/ijchm-06–2015–0323

Li, X., Banning-Taylor, M., Alexander, P. B., & Piction, C. (2014). Demands and constraints on Chinese elderly people traveling abroad: Based on a market survey of Middle-aged and elderly people in Beijing. *Tourism Tribute, 29*(9), 35–43.

Mannheim, K. (1952). *The power of generations: Essays on the sociology of knowledge.* London: Routledge & Kegan Paul.

Nasolomampionona, R. F. (2014). Profile of Chinese outbound tourists: Characteristics and expenditures. *American Journal of Tourism Management, 3*(1), 17–30. http://article.sapub.org/10.5923.j.tourism.20140301.03.html

Prayag, G., Disegna, M., Cohen, S. A., & Yan, H. (Gordon). (2015). Segmenting markets by bagged clustering: Young Chinese travelers to Western Europe. *Journal of Travel Research, 54*(2), 234–250. doi:10.1177/0047287513514299

Ravlin, E. C., Liao, Y., Morrell, D. L., Au, K., & Thomas, D. C. (2012). Collectivist orientation and the psychological contract: Mediating effects of creditor exchange ideology. *Journal of International Business Studies, 43*(8), 772–782. doi:10.1057/jibs.2012.17

Schewe, C. D., & Noble, S. M. (2000). Market segmentation by cohorts: The value and validity of cohorts in America and abroad. *Journal of marketing management,* 16(1–3), 129–142.

So, K. K. F., Liu, W., Wang, Y., & Sparks, B. (2016). Service expectations of Chinese outbound tourists. In X. R. Li (Ed.), *Chinese outbound tourism 2.0* (pp. 185–198). Oakville: Apple Academic Press.

Tse, T. S. M., & Hobson, J. S. P. (2008). The forces shaping China's outbound tourism. *Journal of China Tourism Research*, *4*(2), 136–155. doi:10.1080/19388160802279459

United Nations World Tourism Organization. (2014). UNTWO tourism highlights. *UNTWO*. http://dtxtq4w60xqpw.cloudfront. net/sites/all/files/pdf/unwto_highlights14_en.pdf

Weingarten, R. M. (2009). Four generations, one workplace: A Gen X-Y staff nurse's view of team building in the emergency department. *Journal of Emergency Nursing*, *35*(1), 27–30. doi:10.1016/j.jen.2008.02.017

Wright, J. (2005). Coaching mid-life, baby boomer women in the workplace. *Work: Journal of Prevention, Assessment & Rehabilitation*, *25*(2), 179–183.

Xu, H., Wang, K., & Song, Y. M. (2020). Chinese outbound tourism and soft power. *Journal of Policy Research in Tourism, Leisure and Events*, *12*(1), 34–49. doi:10.1080/19407963.2018.1505105

Yang, F. X., & Lau, V. M. C. (2015). "LuXurY" hotel loyalty – a comparison of Chinese Gen X and Y tourists to Macau. *International Journal of Contemporary Hospitality Management*, *27*(7), 1685–1706. doi:10.1108/ijchm-06-2014-0275

Zhong, L., Wu, B., & Morrison, A. M. (2015). Research on China's tourism: A 35-year review and authorship analysis. *International Journal of Tourism Research*, *17*(1), 25–34. doi:10.1002/jtr.1962

Zhu, H., & Deng, F. (2020). How to influence rural tourism intention by risk knowledge during COVID-19 containment in China: Mediating role of risk perception and attitude. *International Journal of Environmental Research and Public Health*, *17*(10), 3514. doi:10.3390/ijerph17103514

4 Development of China's outbound tourism market

A brief history from then to now

Shuaibu Chiroma Hassan and Ibrahim Sani

Introduction

The history of the Chinese outbound tourism shows how its tourism grows and its current decline due to certain factors (corona disruptions). Therefore, the history of the Chinese outbound shows the stages of development of a destination, factors that lead to the growth and those that lead to its decline. Tourism is seen as the acts and processes of spending one's time away from their homes in the pursuance of relaxation, pleasure and recreations while utilizing commercial provision of services, dating back to the antiquity (Walton, 2021). Tourism is identified by Arionesei et al. (2014) as a phenomenon of socio-economic aspects with strong impacts on cultures, economies and on the societies of mankind.

The period of opening up marks the beginning of China's tourism development after the period of isolation (Bi et al., 2016). After the birth of the 'New China' in 1949, its tourism continued to prosper (Zhang, 2013), though during the isolation period, the tourism turnover was very low (Bi et al., 2016). Up till 2013, the tourism was on a normal development track – the domestic being the major one while the outbound was low (Zhang, 2013). The opening-up period saw the development of its inbound tourism by its citizens. The development of its transportation systems, technology and its rapid industrialization reaching the wider world has been the main reasons for the boost of its tourism industry.

The China's tourism industry is playing a major role in providing jobs, promoting cultural exchange and boosting its economic developments (Zhao & Li, 2018). Tourism has been evolved and becomes a means for international relations between countries (Arionesei et al., 2014). As a result, its popular Belt and Road Initiative (BRI) initiated in 2013 will help to boost its tourism industry. It is a lump sum project for a long-term period between China and countries along the Silk Road to help benefit its large population. The project will encourage more Chinese to travel out to the countries, particularly those within the BRI (One Belt One Road Initiative (OBOR)) and other tourists from those countries to visit China as well.

DOI: 10.4324/9781003121329-5

The Silk Road

At the early time, some of the Silk Route sections were just imaginary directions across the trackless desert or Steppe rather than paths – "the majority of states on the Silk Route traded with their nearer neighbors, and travelers were like participants in the relay race stretching a third of the way around the world" (Goeldner & Ritchie, 2012). The 'Silk Road' was coined in the 19th century, but inter – continental trade and cross – country trading existed quite long before (Chan, 2018). "The Road" aspects of the project refer to the previous 'Maritime Silk Road' which connects between East China and Europe through the Southeast Asia, while the 'Belt' aspect refers to the land 'Silk Road Economic Belt' that links China to Europe through the central Asia (Zhou, 2019).

Marco Polo, from Italy, traveled through the Silk Road to China in the 13th century and became the first explorer from the West to provide a popular lasting account of the region (Goeldner & Ritchie, 2012). The travelers, for protection against attacks, formed a caravan of up to 1,000 camels with their armed escorts. The main long route followed was divided into two areas: the economic and political influence. The Chinese merchants or traders used to escort their merchandise beyond the Great Wall up to *Louvan* where they are bartered or sold to Indians, Parthians and Khushans. The goods were then transported to Syria, Persia and to Greek merchants (Goeldner & Ritchie, 2012).

History of tourism in China

China followed different course for its tourism development compared with the majority of the developed countries of the world which conformed with its economic and political features (Zhang, 2013). Tourism and travel are part of the human activities, they evolved from the ancient periods, a unique phenomenon found in some other civilizations of the world. However, the tourism and travel industry or business was later introduced to China just of recent, and as far later than most of the developing and developed nations globally (Zhang, 2013). During the China's periods of isolation, there was low tourism turnover, with two forms of tourists: the Chinese tourists from outside of the Mainland China and the permitted foreign (diplomats or officials) guests who were all treated like the VIPs (Bi et al., 2016).

After the six decades of the birth of the 'new People's Republic of China' in 1949, its tourism development could be grouped into two main stages: (1) the first '30 years' between 1949 and 1978 during which the emphasis was on the inbound tourism, which was regarded as a kind of 'People-to-people diplomatic activities' and (2) a period during the second '30 years' between 1979 and 2009, with the tourism development incorporating the domestic, the outbound and the inbound components, which is regarded as important industry of the China's national economy; since then, the tourism industry became the main backbone of the whole tourism industry of the country (Zhang, 2013). However, during the period of economic growth, tourism has been identified to be a strong tool – a multi-functional strategic industry for environmental, socio-economic, cultural and political needs in the forms of Green Tourism, Red Tourism and Heritage Tourism for urban developments (Bi et al., 2016).

Up till 2013, the China's tourism at large embarked on the normal development track, that is, the domestic type has been the major one, while the outbound and the inbound grow in line with the market demand (Zhang, 2013). The establishment of the OBOR initiative by Xi Jin Pi in 2013 provides China with access to foreign world (market) and a basic plan of investment bringing more cultural, political and economic exchanges between China and other countries (Zhou, 2019). This Belt initiative, hitherto, provides China with opportunities for the development of its tourism industry. The functions of the China's tourism industry became multi-faceted; the cultural and social functions are for the enhancement of the public welfare, while its economic functions remain very crucial as all were introduced and realized (Zhang, 2013). The development of transportation in China makes travel and movements more convenient as well as its economy, making tourism to be among the most important contributor to its economy since the reform period (Yang, 2019).

The China's domestic tourism is the type of tourism and travel activities for leisure purposes of the Chinese residing within the mainland China excluding the business travels by the Chinese, nor tourism and travels between the mainland China and Hong Kong, Taiwan and Macau. The statistics for the domestic tourism had not existed until in the mid-1980s in China until 1985. Since then, the domestic tourism increased tenfold in terms of trips, while in terms of spending, it increased hundredfold roughly (Zhang, 2013). The 'OBOR' initiative aims to facilitate China's entire development along with other countries linked with the road through communication and cooperation (Zhou, 2019). This helps to stimulate and facilitate the entire Chinese tourism industry.

The period of rapid growth

A new leadership in 2003 forwarded a new 'scientific outlook development' and the 'harmonious society' development phase with developed propaganda (Bi et al., 2016). Due to the economic developments and improved standards of living, as well as the cultural and spiritual needs of its people, the Chinese tourism industry, as a new industry, is continuing to develop into a pillar industry among the tertiary industry for the past decades (Lojo & Cànoves, 2015; Zhao & Li, 2018). In the last three decades, China's tourism had witnessed a rapid and sustained development due to economic growth and a closer integration into the wider world (Bi et al., 2016). During the past decade, between years 2001 and 2010, domestic traveling increased from 784 million to 2,103 million, with annual growth of 10.40%, while domestic spending in the period increased from CNY 352.20 billion to CNY 1,258 billion, with annual growth of 13.60% (Zhang, 2013).

Based on its 13th-year Development Plan, China, after its opening up and reforms, achieved a high leap from the status of 'tourism-short country' to a 'tourist power' (Zhao & Li, 2018). With the five years of development plan, the impacts of the OBOR on China's development have increased with benefits from both sides of the cooperation – encompassing wide areas like development, tourism, technology, trade, communication, education, energy, investments and industry (Zhou, 2019). The tourism contribution rate to the economy in 2015 is 10.80%. Meanwhile, China's inbound tourism, domestic and its outbound tourism, has a full prosperous development, and becomes the largest international tourism consumers as well as the largest domestic tourism market making it the world's fourth largest destination (Zhao & Li, 2018). In the 2014 alone, more than 100 million Chinese had crossed their borders for tourism with only 3.50% going to Europe and over 90% traveled to Asian destinations (Lojo & Cànoves, 2015). During this period, China maintains its position as No. 1 top tourism source market and No. 4 top tourists' destination in the year 2014 (Bi et al., 2016).

Rural tourism in China has a long history dating back to Autumn and Spring periods, where the rural areas have been the farming places for the Chinese Agricultural festivals held annually; entertainments and industrialization were the important transformations; as a result, infrastructures and policies were initiated to develop the countryside (Cheng et al., 2018).

However, in terms of the trips and their associate spending, the domestic tourism market is large; with the average spending of the trips per individual is rather low with an annual growth of only 2.90% lower than their inflation rate of the same period (Zhang, 2013). Based on the Travel and Tourism Competitiveness Index of 2015 released by the World Economic Forum (2015), China ranks 17th among the 141 global economies which portrays a significant leap from 28th position in the year 2013 (Bi et al., 2016). Meanwhile, in the period between its "12th Five Year Plan", China's general tourism contributions to social employments were 10.20%, with the domestic tourists reaching 4.40 billion in 2016 – an increase of up to 10.00%. Furthermore, the revenue from its domestic tourism reached 3.39 trillion yuan, that is, 15.20% increase with the number for outbound residents reaching up to 135.13 million – an increase of 5.70%. The revenues for its international tourism were up to 120 billion dollars – an increase of 5.50% (Zhao & Li, 2018).

Chinese tourism market

The population of China is the largest, with about 1.37 billion people mostly affluent, and serves as the good source of tourism market globally (Ooi, 2007) with about 415 millennials with an increased spending power moving up (Folmer et al., 2019). From 2010, China becomes the second largest economy after the United States (Nasolomampiona, 2014). The outbound, inbound and the domestic tourism have a prosperous development; China has been the largest domestic tourism market, the fourth largest tourists' destination and the largest international tourism consumer (Zhao & Li, 2018). Meanwhile, Bi et al. (2016) reveal that China maintains its global position as No. 1 top tourism source market and No. 4 top tourists' destination up till 2014. China is currently a leading outbound tourism market globally in both total expenditure and departures characterized with a rapid growth as well as potential (United Nations World Tourism Organization [UNWTO], 2019). Its growth in double digit of the outbound tourism is a promising trend due to its high incomes, population size low size of passport holders, policies of tax refund and quick visa facilitations (UNWTO, 2019). Most of the outbound travels (70%), apart from those for Macao and Hong Kong, are operated by the travel agencies (Nasolomampiononа, 2014).

China has undergone two 'New Normal' recently; first, it entered its 'new normal' economy denoting new economic era of development (Huang et al., 2019), and second, the new normal of the post COVID-19 from 2020 to date when the economy is recovering from the pandemic shocks. The former has positive impact on its economy, while the latter has some negative impacts that led to decline in tourism and economic activities. The social stratification has some significant influences on the Chinese outbound tourism market – the millennials who are aged between 15 and 34 years are the main dominant group in the tourism markets. The females travel more for the outbound tourism (53%) than their males' counterpart (47%), whereas the age between 35 and 44 years constitutes up to 33% (UNWTO, 2019). Based on the statistics from Malaysia, Singapore and Thailand, 70–80% of the Chinese tourists were men at their 80s and above, while those aged between 25 and 44 years made up to 65% of the entire outbound tourists. The major travel periods for tour groups are main holidays – the Golden Weeks (spring festivals during the Chinese New Year and National Day Holiday with February and January being more popular months (Nasolomampiononа, 2014).

The Chinese tourism industry is expanding rapidly in recent decades; therefore, understanding its outbound tourists' behavior as well as their experiences is vital for responding to them and their emerging trends as new opportunities (Lojo & Cànoves, 2015). The global outbound tourism industry is notably from China and some Asian countries which provides market opportunities to the global markets (Organization for Economic Cooperation and Development [OECD], 2018). The Chinese middle-class population is continuously rising, and for the past ten years into this expansion, the consumer behavior is shifting with bifurcation among themselves (Ho et al., 2019). There are two market segments (the new generation and the old of 55–65 years) among the Chinese tourists (Curran, 2020). During the post-COVID years, another new generation of "sophisticated Chinese tourists" will emerge in the global tourism market. These will be the first generation of the 'Chinese young adults' who are grown up in affluent families in the country and who started traveling (in groups) with their families (Curran, 2020). A segment of the consumers from lower tier cities are continuing to spend money freely without worries about savings for the future or costs. Meanwhile, other consumers from expensive cities like Shanghai, Beijing and Guangzhou are responding to China's economic growth (Ho et al., 2019).

Domestic tourism industry

During the past decade, between 2001 and 2010, China's domestic traveling increased from 784 million to 2,103 million trips, with annual growth of 10.40%, while the domestic spending in the period increased from CNY 352.20 billion with annual growth of 13.60% (Zhang, 2013). However, Zhang further shows that, in terms of the trips and associated spending, the domestic tourism market is large; while the average spending of the trips per individual is rather low, with an annual growth of only 2.90% – lower than their inflation rate of the same period (Zhao & Li, 2018). One of the main strategies for promoting rural tourism in China is the 'Strategic Plan for Vitalisation of Rural Areas 2018–2022', which is issued by the Central Committee of the Communist Party for implementing rural and *agric* tourism as priority in 2018 (Cheng et al., 2018).

The China's industry scale and industrial systems have been expanding continuously which eventually impacts its tourism industry. Its income from the domestic tourism has increased from 86.40 million yuan in the year 1993 to 39.39 billion yuan in 2016 – an increase of about 45.60 times. The Chinese tourism industry has been enjoying continuous improvements in operations of national economy and its economic development. Moreover, the industry is playing vital role for increasing the employment opportunities, promoting the cultural exchange and boosting its economic development (Zhao & Li, 2018). Thus, the activities and influences of other industrial sectors have positive impacts on China's tourism industry.

Meanwhile, domestic tourism has been the driving force for tourism recovery for many destinations depending on the country (UNWTO, 2020). For instance, the results from the Asia and the Pacific were more positive with regard to domestic tourism toward destinations' recovery based on a survey: World (28%), Africa (20%), Americas (19%), Europe (27%) and Middle East (33%) (UNWTO, 2020). This has similar impacts on China's tourism industry, particularly its domestic tourism.

The China's domestic tourism industry is gradually picking up; the World Expo Museum received 60,000 tourists between July and August 2020 consisting of many local families and students (Yaobin, 2020). Rural tourism in China has a long history dating back to *Autumn* and *Spring* periods, where rural areas have been farming places. Agricultural festivals were held annually; as a result, infrastructures were developed for the country side, and other development policies were initiated (Cheng et al., 2018). In terms of domestic tourism, China is the most active tourism region in the world with a domestic travel recovery at 98% year on year during its recent National Holiday in 2020, with 6.6 billion people traveled across the country for the first nine months of the year (Curran, 2020). The most popular destinations for China's domestic tourism are Xi'an, Sanya, Chendu and Guiyang (Yaobin, 2020).

The 223 major tourism spots of Beijing received up to 1.1 million tourists on the 1 October alone, with more than 1,143% increase compared with 1 May Holiday in 2020 (Xingyu, 2021). A Data from Trip.com shows that the number of families that booked tourism products jumped by 400% higher, accounting for 34% of all bookings on Chinese travels, with high individuals' bookings (Yaobin, 2020). Meanwhile, most of the Chinese domestic tourists have been identified to prefer short-distance travels, making the local tourism more popular. The tourists' preferences for destinations with less congestion and those that are not popular in the country have been revealed (Yaobin, 2020), while entertainment and industrialization are seen as important transformations (Cheng et al., 2018). As a result, there is high increase of 159% of domestic tourists from the first quarter to the second who visited the most popular tourism sports in 2020 due to the country's ability to contain the virus (Xingyu, 2021).

Meanwhile, with effective control of the COVID-19 within China, the domestic tourism industry is recovering quickly (Xingyu, 2021). However, most of the tourism experts see the return of the tourists to pre-pandemic levels to occur before 2023, with positive reboot in 2020 (UNWTO, 2020; Xingyu, 2021). In spite of this, there is good recovery in China's tourism industry from 2020 due to its economic resilience and strict measures for containing the pandemic (Xingyu, 2021). As a result, the domestic tourism is making a "W-shaped" recovery as reported by the China Tourism Academy (Xingyu, 2021). Similarly, the *New York Times* predicted that the China's domestic air travels will reach its full recovery by September 2021 as reported by Yaobin (2020).

In the global trade, China is currently consolidating the status of the 'world's dominant trading power' despite the continuous tensions with the US. The market demands and technology as well as the ability to contain the pandemic virus enable China to recapture the record market share in the global total exports by reopening its factories quickly while others are in lockdown (Bloomberg, 2020). This manifests the China's enduring and resilient role in the manufacturing industry, even with the high traffic levied on the Chinese goods by the Trump administration (Bloomberg, 2020). Meanwhile, the first strata of the Chinese people of 55 to 65 years of age who worked harder for their money in the past 30 years, who have the means and time to travel, will now be the new emerging target market among the Chinese tourists (Curran, 2020). As the global population is aging, some new groups of consumers appear which will benefit tourism as they pursue health tourism (OECD, 2018). In another vein, China Tourism Academy (CTA) forecasts that China, in five years' time, will form a large domestic tourism market with yearly ten billion trips with ten billion yuan spending on average (Xingyu, 2021).

Internal drivers of China's tourism development

Improvement of the Environmental/Ecological Attractions – A Policy is implemented and emphasized to protect and promote natural, ecological and environmental resources for tourism, with reference to social and environmental responsibilities for sustainability. This provides framework for reduced human damages to the environment, development of green tourism, eco-tourism and global warming (Zhao & Li, 2018). One of the main strategies for promoting rural tourism in China is the 'Strategic Plan for Vitalisation of Rural Areas' – 2018–2022, which is issued by the Central Committee of the Communist Party for implementing rural and 'agri-tourism' as a priority in 2018 (Cheng et al., 2018).

Improved Economic Development – With its good economic development policies, China's economic growth has positive impacts on individuals, which gives the Chinese spending power for both domestic and outbound tourism. This transforms China from a big tourists' country to a powerful tourists' country (Zhao & Li, 2018). In addition, Jiang et al. (2018) posit the search for 'Authenticity' as a sociological travel motivation for Chinese outbound tourists.

The Chinese outbound tourism industry

The rapid growth of China's outbound and inbound tourism began in the past 30 years due to the economic reform policies of Deng Xiaoping in early 80s which gave rise for opening to culture, middle class, leisure and create curiosity to discover the outside world (Nasolomampionona, 2014). The Chinese outbound tourism benefited various destinations, mostly in the Pacific and Asia as well as long haul destinations globally (Amato, 2018). Between 1994 and 2004, the outbound tourists from Mainland China increased by up to 17% yearly (Keating & Kriz, 2008). The China's modern outbound tourism began to take shape in 1997, maintaining a double digit of over 20% growth (Nasolomampionona, 2014; UNWTO, 2019). From 2012 to date, China continued to be the largest tourism source market globally, with outbound travels of 150 million tourists (+15%), with spending more than 277 million US dollars (UNWTO, 2019). Due to the thirst for knowledge, new culture, shopping and experiences, the Chinese are becoming the world's most travelers (Nasolomampionona, 2014). China's outbound tourism industry is a 'fast-growing contributor' to its economy on one hand and a source of revenues to international destinations on the other (Jiang et al., 2018).

From the beginning of the Dengist era, the Government of China started to reduce its 'tight grip' on travel for the citizens and allowed them to travel abroad. The laws that restrict traveling within China were stimulated by the 'hukou system' – a type of residency booklet that was revised to allow freedom of movement (Nasolomampionona, 2014). China was traditionally an agricultural society and later transformed into modern and industrial one and eventually to a market economy creating a strong middle class (Nasolomampionona, 2014). In the year 1990, China's boarders were officially opened to allow the citizens to travel outside, initially to Macao and Hong Kong for visit to friends and relatives (Nasolomampionona, 2014).

In 1984, the China's State Council gave approval for the establishment of tours organizations to Macao and Hong Kong, which is the prelude to Chinese outbound tourism. Meanwhile, in 1987, the Ministry of Foreign Trade and Economic Cooperation and the China National Tourism Administration allow the city of Dandong to operate a day tour to the city of Sinuiju in Democratic Peoples' Republic of Korea (Nasolomampionona, 2014). The China's outbound tourism is generated mostly in the wealthy areas of the South-East and East – specifically the highly agglomerations urban areas of Yangtze River Delta (Shanghai); Hebei-Tianjin (Beijin); Chengdu-Chonqing; Pearl River Delta (Guangzhou) and other city clusters (UNWTO, 2019). Political, economic and social factors have significant impacts on China's outbound tourism industry development from the macro-economic perspectives.

Internet has been a powerful force for the promotion of China's outbound tourism – online travel websites are important channels for Chinese outbound tourists (Zhao & Li, 2018). The majority of tourism experts forecast an international tourism rebound in the year 2021, particularly in the third quarter, while 20% of them expects that to occur in 2022 (UNWTO, 2020). Meanwhile, the United Nations World Tourism Organization (2020) identifies factors impeding the recovery of the international tourism as follows: travel restrictions, low consumer confidence and the slow containment of the COVID-19 virus.

Behavior of the Chinese tourists and trends

There are two main market segments among the Chinese tourists (the new generation and the old age of 55–65 years); among them, some will be traveling based on their peculiar preferences, motivations and based on their previous travel experiences. Though they are all Chinese, however, their interests in travels will be different and will therefore need to be well prepared for separately (Curran, 2020). One group of Chinese tourists that traveled to Barcelona, Spain, does not want to travel in an organized tour group – they are the younger ones who are more adventurous than the other ones (Lojo & Cànoves, 2015).

One of the current traveling/tourism trends among the Chinese outbound tourists is their desire to travel individually – on their own – not in a group. This emerging trend is gaining popularity among the younger Chinese tourists; they need to travel even more freely and need access freely to information as well as in-depth knowledge about the destination before and during their stay. They need and prefer to administer their time freely, prefer to pick public transport or rent a car for their movement at the destinations as well (Lojo & Cànoves, 2015). In addition, they also need wide range of their tourism experience and like personal interaction with the local people at the destination (Lojo & Cànoves, 2015). Moreover, Chinese outbound tourists enjoy the local cuisine at the tourism destinations as a way of experiencing an important culture of the local people. Similarly, various scholars examine the link between cultural aspects and the destination marketing (Suntikul et al., 2020). However, Chinese tourists prefer their traditional foods at the destinations, though many of them are eager to try the local dishes of the destination visited (Chang & Mak, 2010).

Meanwhile, some of those independent Chinese travelers return with some visitors who had a prior visit to Europe with organized groups and wanted to repeat their visitation experiences independently as well. The following are the main motivations for this category for traveling: (i) pleasure seeking; (ii) professional and business meetings; (iii) shopping; (iv) attractions from the cities and the European cultures as a result of the occidental culture; (v) curiosity for knowledge of the tourism destination gained from their previous visitation (Lojo & Cànoves, 2015). The second category or group of the Chinese tourists is the students who are studying in the European universities traveling within Europe. Both categories have some knowledge of the European cultures in contrast with the traditional Chinese tourists who use to travel in organized groups (Lojo & Cànoves, 2015).

The trends before the 2020 were gradually changing intrinsically the Chinese tourists' behavior – the majority has lost interests in the package tour travels, with hurried visitations to many tourism destinations, each of them consisting of quick stops for Chinese foods, shopping and snap – shots of major sites visited (Curran, 2020). The 'millennials' are the main dominant tourist groups with family bonds and looking to experience the local lifestyles of the host community. Many of the tourists show interests in group tours; however, more prefer independent tour with independence to drive themselves at the destinations (UNWTO, 2019). In addition, they like 'reckless' shopping at the destination due to their disposable incomes and spending power.

Main destinations of the Chinese outbound tourists

The destinations most visited by the Chinese outbound tourists are mostly the nearby countries of the South-East and North-East Asia, Taiwan, Hong Kong and Macao. Emerging tourist destinations are also gaining the Chinese attractions as well (UNWTO, 2019).

Factors/drivers responsible for the growth of the China's tourism

The Chinese outbound tourism has been growing significantly very fast due to certain factors. The Chinese population size with increased high incomes among the citizens is among the leading factors (UNWTO, 2019). The main factor for the development of the China's tourism industry is its resilient and strong economy – its economic growth within the emerging economies leads to the growth of the middle class which creates the desires for them to travel and explore the world (Yaobin, 2020). Political, social and economic factors have significant impacts as well. As a result, many Chinese embark on outbound tourism due to 'high speed' economic development with a growth rate of 10% GDP (Huang et al., 2019). Other factors for the Chinese outbound growth include easy for facilitating visa, good private and public sectors services and easy payments facilitations (UNWTO, 2019). The external pull (direct flights and easy visa policies, use of technology) and push (travel as fashion and increase in wealth) are other contributing factors (UNWTO, 2019).

Another factor which is the bedrock to China's tourism recovery is its ability to contain the COVID-19 virus at early stage (Xingyu, 2021) more than any country, which helps the industry to strive and enable the citizens to travel for leisure. As a result, China's consumer confidence index hit the ten-year high in early 2019 with a continuous growth of consumer spending in 2020 (Ho et al., 2019). Moreover, China's National Holiday – the period during the mid-July – gave the domestic tourism a boost as over 637 million trips were recorded with corresponding revenues of 466 billion yuan. Meanwhile, the total domestic trips were estimated to be 3.40 billion with corresponding revenues of 2.8 billion yuan throughout the year 2020, with 57% and 41% for 2019, respectively (Xingyu, 2021). Lojo and Cànoves (2015) identify the motivations for the Chinese travelers in Spain: (i) cultural heritage attractions; (ii) Football club (Barcelona FCB); (iii) good weather of Barcelona; (iv) beautiful beaches; (v) the blue sky of the country; (vi) feeling that the city is welcoming to the tourists; (vii) honey moon among others. Financial Interventions – The Chinese Government injected more than US$1.07b as quality guarantee deposit to refund to about 29,000 tourism businesses to help alleviate their financial burdens (Xingyu, 2021).

Discounts and Free Ticket policies – Due to the impacts of the pandemic, discounts and free tickets were given for large number of tourism spots in order to attract more visitors, which stimulate spending on food, accommodation, shopping and transportations (Xingyu, 2021).

The New Silk Road (China Belt and Road Initiative) – The Chinese BRI Network is a great 'lump sum' project – the 'One Belt, One Road' 2013 initiative of China's President Xi Jinping with more than one trillion US dollar investment for infrastructure and constructions development known as Silk Road Economic Belt (Petersmann, 2018). According to Jie (2018), the BRI has three main dimensions: people, trade and capital; the first two have a direct link to tourism while the capital is not directly related. The Project (BRI) connects 88 countries with more than 100 project investments on transport, tourism, infrastructures, energy and education. The network will connect China with Africa, ASEAN Countries, Europe, Middle East and Latin America to promote infrastructure facilities, international co-ordinations on policy, trade, connectivity, human and financial integrations for more than 65 countries neighboring the Silk Road (Petersmann, 2018). This project will have great impact on international tourism industry: China's domestic, inbound as well as outbound tourism. It will also serve as a leverage to China's diplomacy, soft power and economic advantage at the global stage. Eventually, this will be a game changer for China's position in the world. As a result, tourism will boost in those regions and continents with the BRI network which will benefit nearly all the countries.

Activities of the Chinese tourists

The main activities engaged by the Chinese outbound tourists include local gastronomy, shopping, elements of cultural heritage and the exploration of the city. Meanwhile, the main items shopped by the Chinese outbound tourists include the jewelries (Lojo & Cànoves, 2015). Cycling and driving cars, cultural entertainments and adventuring among others. They go to festivals, markets, enjoy street life and the less exploited and unpopular attractions are increasingly becoming more preferred attractions than the popular ones (Amato, 2018).

Discussion

Tourism in China has seen a steady growth from the opening up to date, except for the COVID-19 period in 2020 where it makes a sharp decline due to global lockdowns. However, with good safety measures and the availability of vaccines, China has managed the pandemic leading toward its quick recovery. The Chinese tourism industry in general and outbound tourism developments take the shape of the Butler's Area Life Cycle Model (Singh, 2011), except for the 2020 as a result of the pandemic, though it is now picking up to the 'New Normal'. It has started from the Exploration (during the opening up period), to the Involvement (during the 1990s), to the Development (during the 2000) and the Consolidation Phase (during the 2018) without reaching the Stagnation – the COVID-19 appeared. And now, there is a sharp decline that leads to a V-shape curve. However, with good recovery of the China's domestic tourism industry and its economy due to its economic resilience, and the BRI network taking shape, the Chinese outbound tourism will likely bounce back to better position. The China's inbound tourism will also grow due to many outbound tourists that will visit China for trade, education and tourism in the near future.

Conclusion

The China's tourism industry development follows the Butler's Tourism Life Cycle Model (Singh, 2011), though it is still on its growth life cycle, disrupted by the COVID-19 pandemic creating a V-shape on the curve. The tourism industry is picking up due to China's resilience and its tourism demand as a major global tourism source market. China has an ancient history of travels; many countries traded with China through the famous Silk Road leading to Middle East and Europe. Its modern tourism development started with its opening up to the outside world after its reforms of the Revolution of the 1950s. From then, there was a steady growth in the Chinese tourism industry particularly due to its economic and social policies after the reforms of the Revolution.

The next phase of the China's tourism development continued due to its economic prosperity – as a second largest economy with much production industries, progress in education industry as well as its growing influences on the global trade and politics. This helps to improve the living standards of the Chinese, with their large population and eventually improved its GDP which provides them with disposable income to explore the outside world. Finally, with the initiation of the OBOR or the Silk Road, more direct foreign investments in other countries provide China with economic, social and political power at the global stage in the form of leverage that benefit its tourism industry. However, this is not without a setback, as the West is fighting and challenging the growth of China that creates a form of trade war in recent years. To remain competitive in the global stage, Chinese tourism industry should employ more technology, innovations, new marketing approaches and effective strategies to attract outbound tourist in order to be more competitive as a tourism destination.

References

Amato, R. (2018). *Tourism in focus – China tourism 2018*. Brussels: European Commission. https://www.partnershipsineuropeantourism.eu/wp-content/uploads/2020/06/China-in-Focus-2018.pdf

Arionesei, G., Stanciu, P., Morosan, A.-A., & Cosma, M. R. (2014). Tourism today: Why is it a global Phenomenon?. In *International Conference Sustainable Development in Conditions of Economic Instability* (pp. 240–248). https://www.researchgate.net/publication/267854905

Bi, L., Vanneste, D., & Borg, J. V. D. (2016). Tourism development in China since 1949: A regime perspective. *International Journal of Tourism Policy*, *6*(3–4), 321–340. doi:10.1504/IJTP.2016.10002459

Chan, M. H. T. (2018). The belt and road initiative – the new silk road: A research agenda. *Journal of Contemporary East Asia Studies*, *7*(2), 104–123. doi:10.1080/24761028.2019.1580407

Chang, R. C. Y., & Mak, A. (2010). Food preferences of Chinese tourists. *Annals of Tourism Research*, *37*(4), 989–1011. doi:10.1016/j.annals.2010.03.007

Cheng, Z., Wang, C., & Pu, A. (2018). On the origin, development and prospect of rural tourism in China. *American Journal of Industrial and Business Management*, *8*(12), 2421–2427. doi:10.4236/ajibm.2018.812162

Curran, C. (2020, November). *China offers hope to the global tourism industry* [Post]. LinkedIn. https://www.linkedin.com/pulse/china-offers-hope-global-tourism-industry-caragh-curran-mba/

Folmer, A., Tengxiage, A. (Tanya), Kadijk, H., & Wright, A. J. (2019). Exploring Chinese millennials' experiential and transformative travel: a case study of mountain bikers in Tibet. *Journal of Tourism Futures*, *5*(2), 142–156. doi:10.1108/jtf-02–2019-0018

Goeldner, C. R., & Ritchie, J. R. (2012). *Tourism: Principles, practices, philosophies.* Hoboken, NJ: John Wiley.

Ho, J., Poh, F., Zhou, J., & Zipser, D. (2019, December 18). *China consumer report 2020 – The many faces of the Chinese consumer.* McKinsey & Company. https://www.mckinsey.com/featured-insights/china/china-consumer-report-2020-the-many-faces-of-the-chinese-consumer

Huang, S. (Sam), Li, S., & Wen, J. (2019). Facing the new normal: Special issue on tourism economics in China. *Tourism Economics*, *26*(2), 205–210. doi:10.1177/1354816619879345

Jiang, S., Scott, N., & Ding, P. (2018). Motivations of experienced leisure travellers: A means-end chain study on the Chinese outbound market. *Journal of Vacation Marketing*, 25, 135676671876369. doi:10.1177/1356766718763694

Jie, Y. (2018). China and Southeast Asia: Many belts and roads to turn. In M. Cox (Ed.), *China's belt and road initiative (BRI) and Southeast Asia* (p. 3). Kuala Lumpur: CIMB ASEAN Research Institute. https://www.lse.ac.uk/ideas/Assets/Documents/reports/LSE-IDEAS-China-SEA-BRI.pdf

Keating, B., & Kriz, A. (2008). Outbound tourism from China: Literature review and research agenda. *Journal of Hospitality and Tourism Management*, *15*(1), 32–41. doi:10.1375/jhtm.15.1.32

Lojo, A., & Cànoves, G. (2015). New tourism trends in Barcelona. Chinese tourist experiences and local perceptions. In *Proceedings of the International Conference on Tourism between Spain and China ICTCHS2015, Tianjin, China* (pp. 3–6).

Nasolomampionona, R. F. (2014). Profile of Chinese outbound tourists: Characteristics and expenditures. *American Journal of Tourism Management*, *3*(1), 17–31. doi:10.5923/j.tourism.20140301.03

Ooi, C. S. (2007). *Chinese tourists in Denmark.* Frederiksberg: Copenhagen Business School, Department of International Economics and Management. https://core.ac.uk/download/pdf/17277045.pdf

Organisation for Economic Co-operation and Development. (2018). *OECD tourism and policies 2018.* Paris: OECD Publishing. https://www.oecd-ilibrary.org/docserver/tour-2018-en.pdf?expires=1634492671&id=id&accname=guest&checksum=17868A5C2E237B3997C75249DB51996F

Petersmann, E. U. (2018). Trade and investment adjudication involving 'silk road projects': Legal methodology challenges. *EUI Department of Law Research Paper* (2018/02).

Singh, S. (2011). The tourism area 'life cycle': A clarification. *Annals of Tourism Research*, *38*(3), 1185–1187. doi:10.1016/j.annals.2011.03.005

Suntikul, W., Pratt, S., & Chang, Y. W. J. (2020). Factors that influence Chinese outbound tourists' intention to consume local food. *Journal of Chinese Tourism R*esearch, *16*(2), 230–247. doi:10.1080/19388160.2019.1625841

United Nations World Tourism Organization. (2019). *Guidelines for success in the Chinese outbound tourism market.* Madrid: World Tourism Organisation. https://www.e-unwto.org/doi/epdf/10.18111/9789284421138

United Nations World Tourism Organization. (2020). *UNWTO panel of experts October edition: Return to 2019 levels expected by 2023.* https://www.unwto.org/node/11557

Walton, J. K. (2021, September 28). Tourism. *Britannica*. Retrieved October 19, 2020, from https://www.britannica.com/topic/tourism

Xingyu, C. (2021). China's tourism in review: Rising from the ashes of COVID-19. *CGTN*. Retrieved January 1, 2021, from https://news.cgtn.com/news/2021-01-01/2020-China-s-tourism-in-review-Rising-from-the-ashes-of-COVID-19-WGG3GesW6A/index.html

Yang, K. (2019). Tourism development in China. *A technical report*. doi:10.13140/RG.2.2.12432.76803

Yaobin, H. (2020, June 15). ICH food festival shows off 1,100 – Plus old – Timey brands on digital platform. *CGTN*. Retrieved on January 16, 2021, from https://news.cgtn.com/news/2020-06-15/2020-ICH-Food-Festival-sees-digitalization-of-old-timey-food-brands-RkS8jmKORi/index.html

Zhang, G. (2013) Domestic tourism in China: Policy, practice and prospects. In World Tourism Organization (Ed.), *Domestic tourism in Asia and the Pacific* (p. 55). Madrid: World Tourism Organisation. https://www.e-unwto.org/doi/book/10.18111/9789284414833

Zhao, L., & Li, L. (2018). Analysis of development environment and development trend of Chinese tourism. *IOP Conference Series: Materials Science and Engineering*, *301*, 012101. doi:10.1088/1757–899x/301/1/012101

Zhou, B. (2019). Cooperation between China and Africa under the One Belt One Road Initiative: China's benefits and problems. *Chinese Studies*, *8*(2), 27–41. doi:10.4236/chnstd.2019.82003

5 Development of China's outbound tourism

New trends and travel patterns

Diego Quer and Jianrong Peng

Introduction

In 2004, the World Tourism Organization forecasted that 100 million Chinese tourists would travel abroad by 2020 (United Nations World Tourism Organization [UNWTO], 2004). In 2014, six years before that expected date, the number of outbound trips by Chinese citizens reached that figure, turning China into an engine for the growth of the global tourism market. Nowadays, China is the largest source of outbound tourism in the world both in terms of departures and expenditure (UNWTO, 2020). According to China's Ministry of Culture and Tourism, 169.21 million outbound trips were made by Chinese citizens in 2019 (National Bureau of Statistics of China, 2020b).

The growth of Chinese outbound tourism market is drawing the attention of international tourism researchers from diverse disciplines. In addition, this research topic has been addressed by several literature reviews (Cai et al., 2008; Jin & Wang, 2016; Jørgensen et al., 2017; Keating & Kriz, 2008; Keating et al., 2015; Law et al., 2016; Lojo et al., 2019; Tse, 2015). These literature reviews report that China's outbound tourism market has evolved since the early years of the 21st century. The accumulated experience in traveling abroad, the ever-increasing use of new technologies and smartphones that characterizes China's society, and the emergence of new segments with specific needs are some of the underlying factors behind an evolution that is changing the traditional profile of the Chinese outbound tourist. The aim of this chapter is to provide an overview of the evolution of China's outbound tourism over recent years as well as an analysis of the new trends and travel patterns that are emerging in this market.

China's outbound tourism: Flows and main destinations

Over the last decade, the flows of Chinese outbound tourists have been rising continuously. While in 2010, less than 58 million Chinese citizens traveled abroad; in 2019, that number has tripled, reaching over 169 million. Figure 5.1 reports the evolution of Chinese outbound tourism flows during that period.

DOI: 10.4324/9781003121329-6

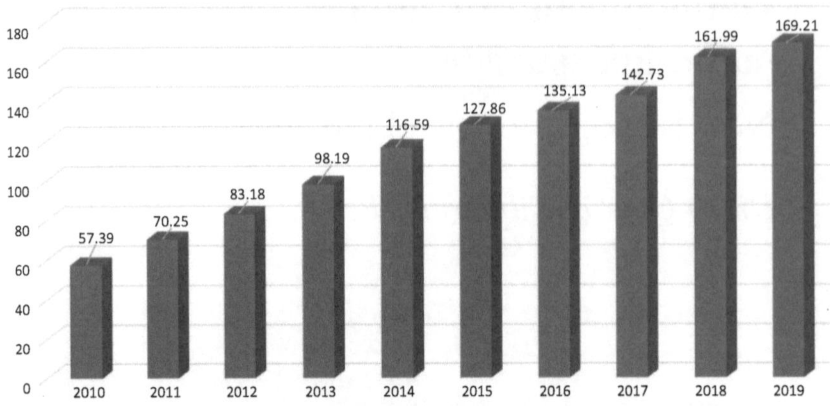

Figure 5.1 Number of Chinese outbound tourists, 2010–2019 (million)
Source: National Bureau of Statistics of China (2020a)

As more Chinese tourists travel abroad year by year and their personal disposable income increases simultaneously, their total spending has been growing steadily (Travel China Guide, 2020). With a total expenditure of USD 254.6 billion in 2019, China is the world's largest tourism spender, representing one-fifth of the international tourism spending (UNWTO, 2017, 2019, 2020).

With regard to destinations, excluding Hong Kong and Macao (special administrative regions of China) and Taiwan, the number of Chinese tourists traveling to overseas destinations exceeded 68 million in 2018, accounting for over 40% of the total (UNWTO, 2019). The top ten destinations for Chinese outbound travelers in 2018 were: Thailand, Japan, Vietnam, Republic of Korea, United States, Singapore, Malaysia, Cambodia, Russia, and Indonesia (Figure 5.2).

Over recent years, more Chinese travelers are attracted by the novelty of emerging tourist destinations. Data from Mafengwo.com (a popular Chinese online travel community) and Trip.com (the largest Chinese online travel agency (OTA), formerly known as Ctrip) reveal that destinations such as Argentina, Bahamas, Brazil, Croatia, Cyprus, Czech Republic, Georgia, Mexico, Montenegro, Serbia, and Turkey are on the rise (UNWTO, 2019).

Beyond the figures, the impressive growth of China's outbound tourism should be analyzed with caution, as Asia still accounts for near 75% of total Chinese outbound trips and Europe ranking second with over 11% (European Commission, 2017). Despite this, European countries are increasingly becoming popular long-haul destinations for Chinese tourists. A joint study by China Outbound Tourism Research Institute (COTRI) and Trip.com estimates that over 12 million of Chinese citizens visited Europe in 2017 (Arlt, 2018). As for specific destinations, based on online bookings data, there are

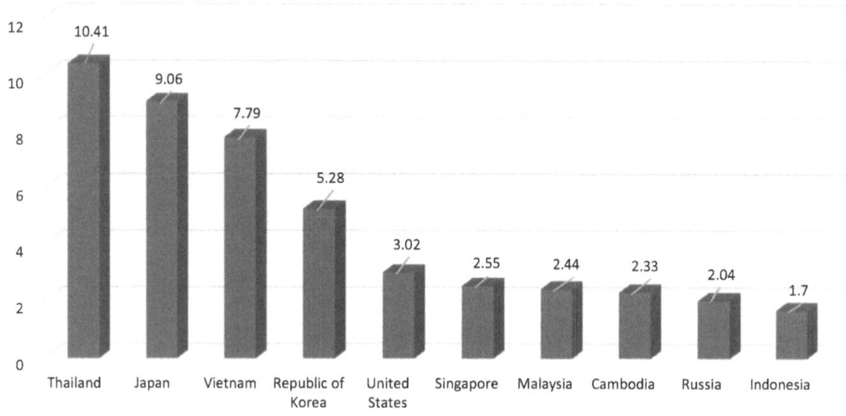

Figure 5.2 Chinese outbound tourists to top ten destinations, 2018 (million)
Source: UNWTO (2019), based on China Tourism Academy/Data Center of Ministry of Culture and Tourism

three European countries among the 20 most popular destinations for Chinese travelers, namely, Italy, Germany, and the UK; moreover, four out of the ten popular destinations with the fastest growth rate in terms of Chinese tourist visits are European: Germany, Spain, Italy, and France (Ctrip and China Tourism Academy, 2018).

Chinese tourism in Europe: Opportunities and challenges

As mature tourist destinations, European countries are focusing on non-traditional markets to ensure a faster growth. Because of the good prospects, an increasing number of countries are targeting China in the last decade (Zhang, 2011). Several factors are expected to boost Chinese outbound tourism in the coming years (European Commission, 2017; Week in China, 2018): a growing middle class in China with increasing disposable income; new attitudes toward leisure among Chinese people as traveling is considered an important part of their experience-oriented lifestyle; the spread of wealth to central and western regions of China; more international flight connections; few visa restrictions; and more Chinese-friendly destinations.

As a result, COTRI forecasts that half of the additional outbound travelers in the next decade will be Chinese, with 390 million outbound trips originating in mainland China by 2030 (Arlt, 2018). The growth potential of the Chinese market is reinforced by the fact that today, only 10% of China's population owns a passport and less than 10% are currently traveling abroad (Ctrip and China Tourism Academy, 2018).

Apart from the potential market growth, there are also two additional characteristics of China's outbound tourism that represent opportunities for

European destinations. First, as pointed out before, China is leading global outbound tourism in terms of expenditure, each Chinese traveler spending on average more per trip than tourists from any other country (UNWTO, 2017, 2019, 2020). Second, Chinese travelers represent a useful tool for fighting against seasonality since they may travel outside the traditional western holiday periods. This is because most Chinese use to travel during the so-called Golden Week holidays (Huang & Wei, 2018): Chinese Lunar New Year (between January and February) and National Day (around October 1).

Chinese tourists' image of Europe

International tourist destinations should take advantage of the image that Chinese tourists have of them. Europe has huge potential to attract Chinese tourists because it may offer not only historical and cultural richness but also opportunities for other types of customized trips as nature, food, sport, and shopping (Andreu et al., 2013). In addition, a high cultural distance may even be a factor attracting Chinese tourists who are looking for new experiences in a different cultural environment (Keating & Kriz, 2008).

Accordingly, past research found that a high cultural distance from China and a large number of World Heritage Sites are destination attributes positively associated with the flows of Chinese tourists visiting each European country (Andreu et al., 2014). Besides, cultural distance moderates the influence of some destination characteristics on the number of Chinese tourist arrivals. Thus, empirical evidence reports that Chinese tourists travel more to destinations with a high level of happiness, but as cultural distance increases, the influence of happiness level gradually decreases (Huang et al., 2021).

Furthermore, a recent analysis of topics discussed in the most popular Chinese social media platforms reveals that Chinese travelers' view of Europe is positive, but there are also some drawbacks (European Commission, 2017). The most frequently mentioned positive characteristics of European destinations are rich culture and history, beautiful and unpolluted environments, many photo opportunities, great shopping, and vibrant street life and cultural activities. The main barriers and annoyances are language and communication problems, unsafety, and difficulties in getting around. Additionally, as we will elaborate later, the new trends of China's outbound tourism market require further adaptation and customization efforts.

Political role of Chinese tourism

It is also worth mentioning that the tourism industry plays a key role in the current China's Government policy agenda. Although the state lets market forces exert influence, it maintains the ultimate control and does not hesitate to intervene to correct the Chinese outbound tourism development path because of its political significance (Tse & Hobson, 2008). Hence, in March 2018, the former Ministry of Culture and China National Tourism

Administration were merged into a new Ministry of Culture and Tourism, with the aim of coordinating the development of cultural and tourism industries, promoting China's soft power and cultural influence, and boosting cultural exchanges with other countries (Arlt, 2018).

Diplomacy is also important. Many countries are streamlining visa procedures to facilitate the entry of Chinese tourists (Ctrip and China Tourism Academy, 2018). Besides, government official visits paid by Chinese President Xi Jinping to other countries contribute to their promotion as tourist destinations among Chinese citizens (Week in China, 2018). In addition, there are recent cooperation agreements between tourism institutions from China and other countries that also matter. For instance, Beijing is the home base of the World Tourism Cities Federation formed by famous tourism cities and tourism-related institutions in the world. Furthermore, the European Union-China Tourism Year was organized in 2018, with the aim of boosting bilateral tourism relations.

In 2013, China launched the Belt and Road Initiative (BRI), an ambitious plan to promote international cooperation between countries that is driving China's diplomacy (Li et al., 2019). Although the tourism industry was not initially included in the BRI, it covers significant projects that may boost the development of tourism in involved countries (Daye et al., 2020). Moreover, several BRI countries are facilitating visa requirements to attract an increasing number of Chinese tourists (Deng et al., 2019; UNWTO, 2019). Up to now, most of the European countries that have joined the BRI are those of Central and Eastern Europe, but there are also others like Italy, Greece, and Portugal, which have signed cooperation agreements under the umbrella of the BRI.

The role played by Chinese tourism companies

In recent years, Chinese companies are becoming key global players, tourism being one of their target industries. There are outstanding examples of Chinese firms that are now controlling important European companies in several tourism fields: Jin Jiang Hotels took over Louvre Hotels and is a significant shareholder in Accor Hotels both based in France; Fosun acquired a majority stake in Club Med (a French vacation village company) and has revived the former UK tour operator Thomas Cook as an OTA; and Trip.com now owns the UK travel metasearch engines Skyscanner and Travelfusion. Apart from allowing the access to managerial know-how, cross-border acquisitions by Chinese firms contribute to boost Chinese tourism flows to target destinations (Quer & Peng, 2019).

Profile of Chinese outbound tourists

The 'first wave' of Chinese outbound tourists during the early years of the 21st century still showed a traditional profile characterized by the following features (Andreu et al., 2013): leisure trips' length between 10 and 15 days visiting several countries in the same trip; enjoying cultural sites and shopping, particularly luxury articles and brand names; liking for Chinese-friendly hotels providing them with Chinese food, a kettle or hot water point, Chinese TV channels, etc.; and a strong preference for traveling in organized groups. This traditional preference for package tours derived from their lack of experience in traveling abroad and also because it was a cheaper and easier way for doing so. Furthermore, state-owned Chinese travel agencies were dominating the market with exclusive rights to arrange tours (Week in China, 2018). Over recent years, the Chinese outbound tourism market is evolving. Today, there is not a single profile of the Chinese tourist, and the following new trends can be identified.

Second-tier cities

Because of the imbalanced economic development in the country, China's outbound tourism is mostly generated in east and south-east coastal areas, including cities like Shanghai, Beijing, Guangzhou, and Shenzhen (UNWTO, 2019). Although these first-tier cities are still the main home base of Chinese outbound tourists, second-tier cities like Chengdu in southwestern China are on the rise. The increasing per capita income of these regions as well as the opening of new international air connections from second-tier cities are behind this trend (Arlt, 2018; Hernández, 2018). Data from Trip.com show that the top ten departure cities for Chinese outbound tourists are Shanghai, Beijing, Chengdu, Guangzhou, Shenzhen, Hangzhou, Nanjing, Wuhan, Tianjin, and Xi'an (Ctrip and China Tourism Academy, 2018).

Online connected

As stated above, high-tech and smartphone use is an idiosyncratic feature of Chinese society (O' Regan & Chang, 2015). Chinese consumers are increasingly connected, using mobile phones for everything. WeChat and Weibo (social media channels), Trip.com (OTA), Mafengwo.com (online travel community), Qiongyou (content platform), Zhihu (social Q&A site), and Huajiao (live streaming platform), among others, are now essential tools for Chinese tourists when looking for information, sharing travel moments, and booking tourism services.

In particular, payments through mobile devices are becoming an intrinsic part of Chinese consumers' culture (Huang et al., 2020). Therefore, Chinese outbound tourists will appreciate not only that China's UnionPay credit cards are accepted by tourism service providers but also that WeChat Pay and Alipay (the most popular mobile payment platforms in China) can be used during their trips abroad. As a result, accepting Chinese-friendly forms of payment is no longer an option, but a necessity for those destinations receiving Chinese tourists (Arlt, 2018; Hernández, 2018).

In addition, Chinese tourists are now very open to the concept of smart tourism, embracing the use of technology and taking advantage of new channels of information services (Lojo, 2020). Consequently, most tour operators and OTAs have launched mobile applications offering a wide range of online services, including bookings, payments, comments posting, etc. Moreover, user-generated content applications such as Douyin (TikTok), the most popular short video sharing platform among young Chinese, are looking at Chinese outbound tourism with contents showing travel recommendations and tips (UNWTO, 2019).

More self-organization

The 'second wave' of Chinese outbound tourists includes more experienced travelers that do not need tour guides. This new profile of Chinese travelers and consumers arises from the growth of urban middle class in China with more people that now can afford to spend more time and money on having fun. By relying on the above-mentioned online platforms, they may travel self-organized as fully independent travelers (FITs) and modular outbound travelers (MOTs). MOTs book parts of tourism services from specialized online platforms that allow them to create their own modular tour as a unique and personalized trip (Arlt, 2015). In order to grasp this market opportunity, Chinese OTAs such as Trip.com are customizing trips, helping Chinese tourists to rent a car and arrange walking tours in host destinations and even to hire locals as tour guides (European Commission, 2017; UNWTO, 2019).

Self-organized traveling is reinforced by the use of private cars to enjoy self-drive holidays. Thus, the use of recreational vehicles is becoming increasingly popular among Chinese tourists, even when traveling abroad to destinations like Australia (Wu & Pearce, 2017; Wu et al., 2018).

Furthermore, Chinese outbound tourists are changing their viewpoint from quantity to quality. Hence, instead of visiting more than three countries per trip at a cheap price engaging only in shopping and sightseeing (as package tour travelers), Chinese FITs and MOTs tend to visit one or two countries per trip and are increasingly interested in new offers based on local culture and nature (Arlt, 2018). This is opening opportunities for new forms of tourism among Chinese travelers, such as slow tourism, namely, a type of tourism focusing on local populations, longer lengths of stay, and more fulfilling tourist experiences (Wen et al., 2021).

In spite of this, package group tours still matter, especially for less young tourists who travel abroad for the first time and/or to destinations that do not consider sufficiently prepared (Ctrip and China Tourism Academy, 2018). However, with more accumulated overseas travel experience and an increasing number of young Chinese traveling abroad, the trend toward travel independently is becoming more pronounced. Recent surveys by the China Tourism Academy (Data Center of Ministry of Culture and Tourism) report that 51% of potential Chinese outbound tourists are now willing to participate in a group tour, whereas in 2017, this percent was 72% (UNWTO, 2019).

Further segmentation and new travel motivations

The new trend of customized trips is opening opportunities for further segmentation of China's outbound tourism market. These are some outstanding examples of segments (Arlt, 2018; Ctrip and China Tourism Academy, 2018; Week in China, 2018): foodies or *chihuo* (the intrinsic passion of Chinese citizens for food is leading some of them to try and learn about local food in countries such as Italy, France, and Spain); culture lovers (with cultural heritage interests linked to arts, literature, and architecture); shoppers (who enjoy buying products, mainly, luxury goods); polar travelers (attracted by natural phenomena such as aurora borealis in North Europe); family travelers (amusement parks for parent–children trips and visiting relatives who study overseas); millennials (younger, adventurous, and independent travelers who like off-the-beaten-track places); silver tourists (elder Chinese who still prefer package tours and look for new experiences such as cruises); and women tourists (who increasingly travel abroad on their own will, attracted by movie location tours among other motivations).

Each segment has different travel motivations and preferences. As a consequence, recent studies are focusing on specific segments of Chinese outbound tourists, trying to investigate their distinctive behavioral patterns. Thus, Li et al. (2017) distinguished three groups of Chinese outbound travelers: 'journey beginners' (usually young people, sensitive to price, and without previous experience in foreign travel); 'conspicuous consumers' (middle-aged or young people, some of them with prior experience in long-haul outbound travel, who seek high-quality and somewhat expensive package tours); and 'prestige pursuers' (usually old people who appreciate the fame of the host country and prefer to visit well-known destinations among Chinese citizens and want to see as many attractions as possible).

Similarly, Lojo and Li (2018) reported differences among segments of mainland Chinese tourists in Spain: while independent travelers complained about Chinese language services, package tour travelers were less satisfied with the trip and the destination in general; moreover, specific destination attractions, like natural resources and beaches, were motivations for repeat visitors, but less decisive for first-time visitors. The heterogeneity of Chinese outbound tourism market may also represent an advantage for less-visited tourist destinations. Hence, Jørgensen et al. (2018) found that more-experienced Chinese tourists tend to diversify and change their behavior, one of the newest trends being explored in second-tier destinations abroad.

Among these segments of Chinese outbound tourists, the so-called millennials or Generation Y deserve special attention. They represent a new generation of younger and more adventurous travelers born in the 1980s and 1990s that today account for approximately 400 million people in China (Week in China, 2018). Besides, it is possible to distinguish two categories of Chinese millennials, namely, *balinghou* (post-1980s) and *jiulinghou* (post-1990s), the latter often said to be more open-minded, rebellious, and individualistic (Moore & Chang, 2014). Millennials are more experienced consumers, spending more than older generations on activities like eating out, doing sports, and traveling, this being a part of their experience-oriented lifestyle (European Commission, 2017).

In addition, this new generation of Chinese millennial tourists shows a distinctive behavioral pattern from earlier generations (Dichter et al., 2018; Week in China, 2018): they are better prepared to travel independently and readier to book their own accommodation and make travel arrangements; many of them have money to spend and a lower need of saving for their first home (because their parents often help them to buy it); they have more diversified needs for customized products (although they still like shopping, it is no longer a priority when they go on holidays since they prefer to enjoy nature, try local cuisine and carry out leisure activities that offer learning-by-doing experiences); they are more digitally savvy (using mobile apps more often than older generations, although not exclusively, since they also resort to OTAs for travel booking); and their consumer habits are less predictable (they used to continue searching for new activities during their trips using their smartphones, and they usually change their minds and are more difficult to capture as repeat visitors).

Furthermore, as stated above, Chinese millennial tourists have new motivations to travel that go beyond the traditional drivers of shopping and sightseeing. Millennials show a more active tourism behavior and highly appreciate destination attributes more related to experience-seeking. A recent lifestyle survey among wealthy Chinese millennials shows that over 70% of them have already traveled abroad, around 45% plan between three and five trips a year, and consider safety more important than shopping (Rambourg, 2018).

An analysis of young Chinese tourists in Italy revealed that they were more interested in taking part in the Italian daily life, namely, living "like their fellow hosts" or "living a destination from the inside" (Antonioli Corigliano, 2011). Similarly, empirical evidence from Chinese tourists visiting Spain shows that younger tourists (18–29 years old) prefer independent trips and have a wider range of motivations for activities at the destination that go beyond sightseeing and monumental heritage as they include learning, relaxation, sun and beaches, nature, cultural and local festivities, history, wine, and food, among others (Lojo, 2020).

It is estimated that around 70% of current Chinese outbound travelers are millennials, and given that 74 million of them are set to graduate at Chinese universities over the next decade, it is expected that they will travel overseas in large numbers since they are well informed and language barriers are lower for them compared to their parents (Kawano et al., 2015).

Conclusion

This chapter has depicted an overview of the recent evolution of China's outbound tourism, highlighting its new trends and travel patterns. We have also provided some insights regarding the opportunities and challenges that this huge and evolving market represent for tourist destinations that are in the maturity stage. We conclude that the new trends are reshaping travel behavior of Chinese tourists, and this, in turn, requires further segmentation. Among the new segments of China's outbound tourism market, millennials deserve special attention, as they represent a new type of Chinese tourist whose distinctive behavior is demanding the development of tailor-made strategies to meet their specific needs.

Finally, the consequences of the COVID-19 pandemic must also be considered. In the short term, uncertainty and volatility will have a strong impact on Chinese outbound tourism flows. Once travel restrictions in many international destinations are progressively reduced, research efforts should be made to elucidate whether the post-pandemic China's outbound tourism remains as before or not. Some promising avenues for future research could be to analyze the impact of security concerns on Chinese outbound tourists' behavior and the evolution of the above-mentioned trends like independent travel, self-drive holidays, and slow tourism.

References

Andreu, R., Claver, E., & Quer, D. (2013). Chinese outbound tourism: New challenges for European tourism. *Enlightening Tourism: A Pathmaking Journal*, *3*(1), 44–58. http://hdl.handle.net/10272/6796

Andreu, R., Claver, E., & Quer, D. (2014). Destination attributes and Chinese outbound tourism to Europe. *Journal of China Tourism Research*, *10*(3), 275–291. doi:10.1080/19388160.2013.855692

Antonioli Corigliano, M. (2011). The outbound Chinese tourism to Italy: The new graduates' generation. *Journal of China Tourism Research*, *7*(4), 396–410. doi:10.1080/19388160.2011.627015

Arlt, W. G. (2015, December 1). *Chinese modular travelers: Independent but organized*. Forbes. Retrieved October 25, 2021, from https://www.forbes.com/sites/profdrwolfganggarlt/2015/12/01/chinese-modular-travelers-independent-but-organised/?sh=7228c76f6681

Arlt, W. G. (2018). *Customized travels of Chinese visitors to Europe*. China Outbound Tourism Research Institute (COTRI). https://china-outbound.com/wp-content/uploads/2018/11/madridNov2018.pdf

Cai, L. A., Li, M., & Knutson, B. J. (2008). Research on China outbound market: A meta-review. *Journal of Hospitality & Leisure Marketing*, *16*(1–2), 5–20. doi:10.1080/10507050802096778

Ctrip and China Tourism Academy. (2018). 2017 China outbound tourism travel report. Retrieved October 25, 2021, from https://pdffox.com/2017-china-outbound-tourism-travel-report-pdf-free-q-117090-pdf-free.html

Daye, M., Charman, K., Wang, Y., & Suzhikova, B. (2020). Exploring local stakeholders' views on the prospects of China's Belt & Road Initiative on tourism development in Kazakhstan. *Current Issues in Tourism, 23*(15), 1948–1962. doi:10.1080/13683500.2019.1700941

Deng, T., Hu, Y., & Yang, Y. (2019). How geographic, cultural, and institutional distances shape location choices of China's OFDI in tourism? – An empirical study on B&R countries. *Asia Pacific Journal of Tourism Research, 24*(8), 735–749. doi:10.1080/10941665.2019.1630451

Dichter, A., Chen, G., Saxon, S., Yu, J., & Suo, P. (2018). *Chinese tourists: Dispelling the myths.* New York: McKinsey & Company.

European Commission. (2017). *Tourism flows from China to the European Union - Current state and future developments.* Luxembourg: Publications Office of the European Union. https://op.europa.eu/en/publication-detail/-/publication/bda9c8a2-4c43-11e7-a5ca-01aa75ed71a1/language-en/format-PDF

Hernández, E. (2018). Chinese tourism in Spain: Opportunity and new trends. *Elcano Royal Institute*, ARI 58/2018, 1–8. Retrieved October 25, 2021, from http://www.realinstitutoelcano.org/wps/wcm/connect/c877ebd5-6d80-4114-9881-17585dc13d7a/ARI48-2018-Hernandez-Turismo-chino-Espana-oportunidad-nuevas-tendencias.pdf?MOD=AJPERES&CACHEID=c877ebd5-6d80-4114-9881-17585dc13d7a

Huang, S., & Wei, X. (2018). Chinese outbound travel: Understanding the socioeconomic drivers. *International Journal of Tourism Research, 20*(1), 25–37. doi:10.1002/jtr.2150

Huang, Y., Wang, X., & Wang, X. (2020). Mobile payment in China: Practice and its effects. *Asian Economic Papers, 19*(3), 1–18. doi:10.1162/asep_a_00779

Huang, Z., Huang, S., Yang, Y., Tang, Z., Yang, Y., & Zhou, Y. (2021). In pursuit of happiness: Impact of the happiness level of a destination country on Chinese tourists' outbound travel choices. *International Journal of Tourism Research, 23*(5), 713–725. doi:10.1002/jtr.2421

Jin, X., & Wang, Y. (2016). Chinese outbound tourism research: A review. *Journal of Travel Research, 55*(4), 440–453. doi:10.1177/0047287515608504

Jørgensen, M. T., Law, R., & King, B. E. (2017). Understanding the past, anticipating the future – a critical assessment of China outbound tourism research. *Journal of Travel & Tourism Marketing, 34*(7), 880–891. doi:10.1080/10548408.2016.1220889

Jørgensen, M. T., Law, R., & King, B. E. (2018). Beyond the stereotypes: Opportunities in China inbound tourism for second-tier European destinations. *International Journal of Tourism Research, 20*(4), 488–497. doi:10.1002/jtr.2198

Kawano, S., Lu, J., Tsang, R., & Liu, J. (2015, November 20). *The Chinese tourist boom.* Tokyo: Goldman Sachs. https://www.goldmansachs.com/insights/pages/macroeconomic-insights-folder/chinese-tourist-boom/report.pdf

Keating, B., & Kriz, A. (2008). Outbound tourism from China: Literature review and research agenda. *Journal of Hospitality and Tourism Management, 15*(1), 32–41. doi:10.1375/jhtm.15.32

Keating, B. W., Huang, S., Kriz, A., & Heung, V. (2015). A systematic review of the Chinese outbound tourism literature: 1983–2012. *Journal of Travel & Tourism Marketing, 32*(1–2), 2–17. doi:10.1080/10548408.2014.986009

Law, R., Sun, S., Fong, D. K. C., Fong, L. H. N., & Fu, H. (2016). A systematic review of China's outbound tourism research. *International Journal of Contemporary Hospitality Management, 28*(12), 2654–2674. doi:10.1108/IJCHM-06-2015-0323

Li, C. X., McCabe, S., & Chen, Y. (2017). Destination choice of Chinese long-haul outbound tourists and market segmentation. *Journal of China Tourism Research, 13*(3), 298–315. doi:10.1080/19388160.2017.1398118

Li, J., Liu, B., & Qian, G. (2019). The belt and road initiative, cultural friction and ethnicity: Their effects on the export performance of SMEs in China. *Journal of World Business*, *54*(4), 350–359. doi:10.1016/j.jwb.2019.04.004

Lojo, A. (2020). Young Chinese in Europe: Travel behavior and new trends based on evidence from Spain. *Tourism: An International Interdisciplinary Journal*, *68*(1), 7–20. doi:10.37741/t.68.1.1

Lojo, A., & Li, M. (2018). Segmentation by experiential familiarity and travel mode of the Chinese outbound market to Spain. *Journal of China Tourism Research*, *14*(1), 100–121. doi:10.1080/19388160.2018.1442278

Lojo, A., Li, M. & Cànoves, G. (2019). Co-authorship networks and thematic development in Chinese outbound tourism research. *Journal of China Tourism Research*, *15*(3), 295–319. doi:10.1080/19388160.2018.1512433

Moore, R. L., & Chang, Z. (2014, July 15). From *balinghou* to *jiulinghou*, China's millennials come of age. *World Politics Review*. Retrieved October 25, 2021, from https://www.worldpoliticsreview.com/articles/13919/from-balinghou-to-jiulinghou-china-s-millennials-come-of-age

National Bureau of Statistics of China. (2020a). *National data – Tourism - Number of Chinese outbound visitors*. Retrieved October 25, 2021, from https://data.stats.gov.cn/english/easyquery.htm?cn=C01

National Bureau of Statistics of China. (2020b, February 28). *Statistical communiqué of the People's Republic of China on the 2019 national economic and social development*. Retrieved October 25, 2021, from http://www.stats.gov.cn/english/PressRelease/202002/t20200228_1728917.html

O' Regan, M., & Chang, H. (2015). Smartphone adoption amongst Chinese youth during leisure-based tourism: Challenges and opportunities. *Journal of China Tourism Research*, *11*(3), 238–254. doi:10.1080/19388160.2015.1077181

Quer, D., & Peng, X. (2019). Multinacionales de mercados emergentes y adquisiciones internacionales: el caso de las empresas chinas en el sector hotelero español. *Cuadernos de Turismo*, *44*, 467–485. doi:10.6018/turismo.44.404981

Rambourg, E. (2018). *China's luxury lifestyle*. London: HSBC Global Banking and Markets.

Travel China Guide. (2020). *China tourism: Statistics and data*. Travel China Guide. https://www.travelchinaguide.com/tourism/

Tse, T. S. M. (2015). A review of Chinese outbound tourism research and the way forward. *Journal of China Tourism Research*, *11*(1), 1–18. doi:10.1080/19388160.2014.984798

Tse, T. S. M., & Hobson, J. S. P. (2008). The forces shaping China's outbound tourism. *Journal of China Tourism Research*, *4*(2), 136–155. doi:10.1080/19388160802279459

United Nations World Tourism Organization. (2004). *WTO in Asia and the Pacific 2004*. Madrid: UN World Tourism Organization. https://www.e-unwto.org/doi/book/10.18111/9789284408184

United Nations World Tourism Organization. (2017). *Penetrating the Chinese outbound tourism market – Successful practices and solutions*. Madrid: UN World Tourism Organization. https://www.e-unwto.org/doi/book/10.18111/9789284418992

United Nations World Tourism Organization. (2019). *Guidelines for the success in the Chinese outbound tourism market*. Madrid: UN World Tourism Organization. https://www.e-unwto.org/doi/book/10.18111/9789284421138

United Nations World Tourism Organization. (2020). *UNWTO tourism dashboard*. Madrid: UN World Tourism Organization. https://www.unwto.org/country-profile-outbound-tourism

Week in China. (2018, Spring). A billion bookings. *Week in China*. https://www.weekinchina.com/zine/focus-editions/a-billion-bookings/

Wen, J., Kozak, M., Yang, S., & Liu, F. (2021). COVID-19: Potential effects on Chinese citizens' lifestyle and travel. *Tourism Review*, *76*(1), 74–87. doi:10.1108/TR-03-2020-0110

Wu, M. Y., & Pearce, P. L. (2017). Understanding Chinese overseas recreational vehicle tourists: A netnographic and comparative approach. *Journal of Hospitality & Tourism Research*, *41*(6), 696–718. doi:10.1177/1096348014550869

Wu, M. Y., Pearce, P. L., & Li, Q. (2018). Chinese behind the wheel: Factors affecting their satisfaction with international self-drive holidays. *Journal of Destination Marketing & Management*, *9*, 12–19. doi:10.1016/j.jdmm.2017.09.004

Zhang, G. (2011). Analyses and forecasts on China's outbound tourism development 2010–2011. In G. Zhang, R. Song, & D. Liu (Eds.), *Green book of China's tourism 2011. China's tourism development analysis and forecast* (pp. 89–104). Heide: China Outbound Tourism Research Institute (COTRI).

Part II

Travel experiences of Chinese outbound tourists

6 China's long-haul travel experiences

The development of Chinese tourism in Italy

Lala Hu, Mirko Olivieri and Jun Wen

Introduction

In Italy, tourism represents a main industry of the local economy, accounting for 13% of the Italian GDP. In 2020, its market value amounted to over 230 billion euros and employed about four million employees, equal to about 15% of the total workforce (PricewaterhouseCoopers [PwC], 2021). Indeed, Italy is a popular destination thanks to its cultural heritage, quality food, and famous fashion brands. However, the dramatic effects of the pandemic impacted all countries in the world, including Italy. After the COVID-19 outbreak in the first half of 2020, the tourism industry has been strongly affected by travel restrictions at international level and in some cases at national level as well (Baum & Hai, 2020). As a consequence, long-haul travel experiences have been replaced by alternative forms of transport for short-distance trips, favoring domestic destinations (Skift Research, 2020). This trend represents a challenge for Italian tourism companies, which used to rely on the arrival of tourists from distant countries, often outside Europe. During the period January–September 2020, foreign tourists coming to Italy decreased by 68.6% as compared to the same period of the previous year (Istituto Nazionale di Statistica [Istat], 2020). The highest decrease was referred to tourists from China (−90.9%) and the United States (−89.5%) (Gagliardi, 2020). Regarding the Chinese tourism, in the years prior to the COVID-19 pandemic, Italy had become the favorite long-haul destination of Chinese tourists, who amounted to about three million arrivals per year (Eurostat, 2018).

Given its relevance, this phenomenon has attracted the attention of both academics and professionals. The tourism literature has provided a review of the studies on the outbound tourism from China, also by elaborating theoretical frameworks (e.g., Keating & Kriz, 2008). However, the academic research related to long-haul destination choice is still scarce (Masiero et al., 2019).

Hence, the aim of this chapter is to develop a conceptual framework of China's long-haul tourism by focusing on Italy as a main destination. In order to achieve this objective, we reviewed the recent literature on long-haul travels and Chinese tourists' characteristics. Then we analyzed the

DOI: 10.4324/9781003121329-8

evolution of Chinese tourism in Italy by discussing secondary data on tourism flows from China to Italy before the COVID-19. Our contribution is twofold: we attempt to advance tourism literature focusing on Chinese tourists; moreover, we provide managerial implications for international companies operating in such sector. The remainder of this chapter is structured as follows: in the "Long-haul travels" section, we revise the recent literature on long-haul travel; in the "Chinese tourists' characteristics" section, the focus is on the Chinese tourists' characteristics; in the "The Chinese tourism in Italy" section, we analyze the evolution of the Chinese tourism in Italy; in the "China's long-haul tourism: A conceptual framework" section, we develop a conceptual framework of the Chinese long-haul tourism; finally, the conclusion is given in the Conclusion section.

Long-haul travels

Bonetti et al. (2006) postulate the tourism destination as a system endowed with artistic, cultural, natural resources, and a network of actors in tourism activities. In particular, the authors identify four tourism system categories. The first one is the market cluster in which a governing body is not recognized, and collaborative relationships are not established between the players in the system; in this case, factors external to the system determine its evolution. The second type is the district which, like the previous one, does not see the presence of a unitary government. Unlike the market cluster, however, actors try to establish relationships and collaborate to make decisions. Then, there are local tourist systems that have a governing body that takes decisions shared by all the players in the system who have established close relationships. The last type is that of constellations in which the governing body has great power, guides the decision-making process, and determines the evolution of the destination; here, relationships are established between the actors of a hierarchical type.

Scholars observe that the topic of distance in travel is influenced by the profile of the tourist: people who have traveled long-haul tend to be older, with a higher income and more travel experience as well as higher educational qualifications (McKercher, 2008). In other words, the long-haul travel market is more exclusive and upscale. This is also supported by the study by Lo and Lam (2004), who particularly emphasized the role of the travel agency in outbound tourism. The two authors summarized the results of their study by arguing that the travel agencies must consider three "Cs", namely Customer solutions, Communication mix, and Counselors. "Customer solutions" refer to safety, destination, and transport. "Communication mix" concerns the appropriate use of communication and marketing tools in the different phases of the consumer's purchase path. "Counselors" represent the consulting services to be offered to travelers, for example, by responding to their questions on different aspects of the travel experience.

In addition to the key elements of the travel experience described so far, several scholars have also focused on the impact of distance in destination marketing studies (Bao & McKercher, 2008; Lew & McKercher, 2002; McKercher, 2008), arguing that the behavior of long-haul tourists is different from the one of short-haul travelers. Depending on the travel distance, travelers' consumption habits usually change: short-haul markets mainly involve women and young people; while long-haul tourists usually travel with their family or spouses and not friends or business associates (Bao & McKercher, 2008). Consequently, preferences during the stay may also change. Indeed, long-haul travelers prefer visiting cultural heritage, while short-haul tourists are driven by shopping or relax activities (Ho & McKercher, 2014).

Therefore, for marketers, it is relevant to carry out reliable market research and exploit the results to implement effective destination marketing activities (Corigliano, 2011). Furthermore, structured marketing strategies would allow tour operators to provide travelers with more targeted and memorable experiences.

In recent years, long-haul tourism has been strongly affected by the Internet. Thanks to digital platforms, tourists can access various alternatives in their travel decisions. Li et al. (2017) argue that the decision-making process can be represented as a funnel in which many alternatives are eliminated in the initial phase to arrive at a smaller set of considerations. Afterwards, travelers finalize the decision by confronting the alternatives of the consideration set. Previous research has focused on the identification of the attributes which play a key role in choosing travel destinations. For example, according to some scholars, destination image, familiarity, and affection constitute explanatory factors (Fry & Prentice-Dunn, 2005).

Another key characteristic of long-haul tourism is represented by multi-destination travels. This tourists' itinerary involves visiting multiple destinations rather in the same trip than reaching a single focal destination (Yang et al., 2016). Multi-destination trips are preferred to single-destination trip because of the utility derived by combining the characteristics of the various destinations (Tussyadiah et al., 2006). In fact, in long-haul travel experiences, tourists usually plan a multi-destination visit due to two main factors, that is, opportunity costs and travel time (Yang et al., 2016). The literature has widely investigated this phenomenon of multi-destination visits by distinguishing between primary and secondary destinations (Masiero et al., 2019). Based on this categorization, secondary destinations represent interesting places to visit on the way to the main destination. However, the planning processes for visiting primary and secondary destinations are different, as are the ways of consumption and the reasons of the visit.

This phenomenon is confirmed by Chinese tourists, who usually visit multiple countries in their long-haul travels (McKinsey, 2018), especially in Europe. Tourists from China are an increasing contemporary market; however, they are still considered quite new. Therefore, destination organizations might be still unfamiliar with Chinese tourists (Pearce et al., 2015). Other barriers are represented by language and transportation (Gu & Huang, 2018). In the next paragraph, we analyze the main characteristics of Chinese tourists.

Chinese tourists' characteristics

Before the COVID-19 pandemic, China had ranked first as the world's top spender in international tourism for eight consecutive years since 2012. In 2019, the total Chinese outbound expenditure amounted to US$255 billion ahead of US tourists (US$152 billion expenditure) (United Nations World Tourism Organization [UNWTO], 2020).

Following China's economic development, according to Jin et al. (2019), Chinese outbound tourist shopping reflects a new Chinese consumer culture, which combines elements from the Chinese traditional values and Western ones. According to Prayag et al. (2015), the rapid growth of Chinese tourists in recent years is a key cultural phenomenon to be analyzed. In particular, the authors point out that the academic debate on this phenomenon has focused on aspects such as Chinese travelers' perceptions of long-haul travel, purchasing behaviors, and evaluations of their travel experiences.

Chinese tourists are quite diversified with regard to travel motivations and behaviors (Arlt, 2006). For instance, some scholars (Sparks & Pan, 2009) have shown that Chinese younger tourists want to be more autonomous during their travel experiences. Li et al. (2013) focused on Chinese female tourists who prefer to travel independently. According to Paris (2012), independent travel is characterized by greater heterogeneity. The main motivations that lead independent travelers to undertake a new travel experience include the exploration of other cultures, the growth of knowledge, the mental relax, and the willingness to see new things (Prayag et al., 2015).

Li et al. (2013, p. 791) have segmented long-haul Chinese outbound tourists into three groups: (1) Entertainment/Adventure Seekers, who desire fun and entertainment in their travels; (2) Life-seeing Experience/Culture Explorers, who look for unique and cultural experiences; and (3) Relaxation/Knowledge Seekers, who mainly travel to relax and augment their knowledge.

Regarding the destination attributes evaluated by Chinese travelers, Sparks and Pan (2009) have identified elements related to the destination (i.e., natural scenery and quality infrastructure) and other elements related to the tourist's individual and social needs (i.e., autonomy, inspirational motivations, and social self-enhancement). Another main characteristic of Chinese travelers is the use of technology before, during, and after their trips. With 989 million users as of December 2020, China hosts the largest Internet user base in the world (China Internet Network Information Center [CNNIC], 2021). Local netizens use e-commerce and m-commerce (mobile commerce) to buy products basically from any category. Travel (including accommodation) represents the second sector in e-commerce expenditure in China reaching US$ 228.5 billion sales, only behind fashion and beauty industry (We Are Social, 2020).

Chinese tourists like sharing their experiences online on blogs and social media like WeChat and Weibo. Most content display travelers' experiences of attraction and food in the destination (Tse & Hobson, 2008). The abundance of practical information (i.e., suggestions and recommendations) highlights that the main goal of Chinese tourists is represented by informing other users rather than a way to express their own individual identity (Jin et al., 2019). Furthermore, an increasing number of Chinese independent tourists have created travel blogs to share their own travel routes (Tse & Zhang, 2013; Wu & Pearce, 2014). During recent years, Italy has become Chinese travelers' top visited European country. Therefore, it is relevant to understand how the Chinese tourism has evolved and the characteristics of the travel experience in the most preferred long-haul destination – Italy.

The Chinese tourism in Italy

According to academic studies (Lin et al., 2015), when travelers have the opportunity to travel, they first visit their own country and then they travel abroad. Therefore, travelers who move out of their country of origin already have travel experiences, and their behavior aims to acquire new knowledge and encounter other cultures.

Regarding the distribution of Chinese tourists, in 2019, domestic travel accounted for 95.2% of all travels (Statista, 2020c); therefore, outbound tourism is still limited in China despite growing rapidly. Chinese tourists represent an interesting segment among long-haul travelers to be analyzed from an academic perspective (Sparks & Pan, 2009; Xu & McGehee, 2012). As a matter of fact, worldwide spending on international tourism has increased significantly over the past decade, with China overtaking the United States in total tourism spending. The contribution of Chinese travelers in global tourism ranks first in the world. According to Statista (2020b), in 2019, Chinese tourists spent approximately US$254.6 billion traveling overseas, while in the previous year, they spent around 25.8 million nights in Europe. Regarding the number of arrivals in Italy, they sharply increased over the past decade from 716.55 thousand arrivals from China and Hong Kong Chinese tourists in 2008 to approximately 3.2 million arrivals (Statista, 2020a). Since then, the number of Chinese tourists has remained high, placing China as a relevant entry market for Europe.

Among EU-27's countries, Italy ranks as the most favorite destination ahead of France, reaching about 3 million arrivals per year before the COVID-19 pandemic (Eurostat, 2018). From 2008 to 2019, the average number of nights spent by Chinese tourists in Italy increased from 1.22 million

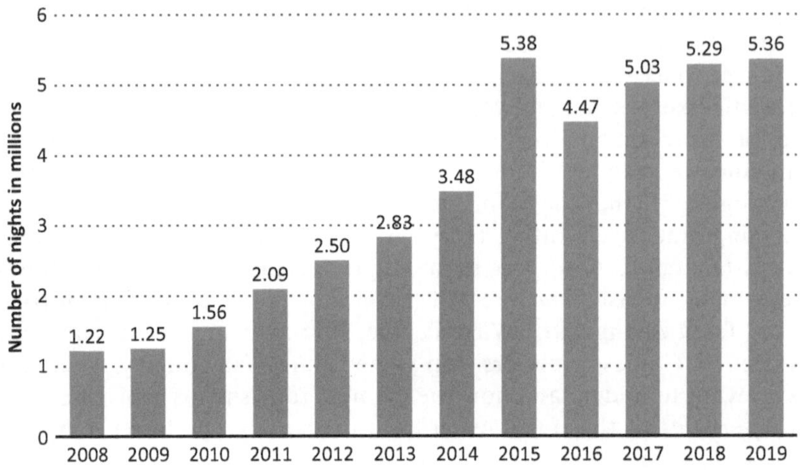

Figure 6.1 Number of overnight stays by tourists from China (including Hong Kong) in travel accommodation in Italy from 2008 to 2019 (in millions)
Source: Statista (2020a)

to 5.36 million (the peak was reached in 2015 with an average of 5.38 million nights) (Statista, 2020b), as shown in Figure 6.1.

Based on our analysis, the main motivations behind the Chinese tourism in Italy are represented by (1) cultural heritage, (2) authenticity seeking, and (3) luxury travel experience. World heritage sites have a relevant tourist-enhancing effect (Yang et al., 2010). In this regard, Italy and China are the countries with the largest number of UNESCO world heritage sites in the world with 58 and 56 world heritage sites, respectively (United Nations Educational, Scientific and Cultural Organization [UNESCO], 2021). Some scholars (Pearce et al., 2013) have found that Chinese people show a strong interest in Italy due to the cultural and historical heritage that they learn at school. The most visited Italian cities by Chinese tourists are Milan, Rome, Florence, and Venice as they represent the Western culture world visually and symbolically (Corigliano, 2011).

Another motivation that has recently increased in the Chinese outbound tourism is represented by authenticity (Zhu et al., 2020), which denotes a higher sophistication. Authenticity seeking leads Chinese tourists to look for experiences that are typical of the destination country such as dining tourism (Cai et al., 2021) and purchase of quality products (Jin et al., 2019).

A more traditional motivation behind Chinese outbound tourism is represented by luxury travel experience. First, it reinforces consumers' social status (Zhang & Tse, 2018). Then, during their trips, Chinese tourists enjoy shopping (Dai et al., 2017), especially for luxury brands. The rising Chinese middle class particularly appreciates exclusive and high-quality products that are often purchased in the country of origin (COO) (Rovai, 2016). At world level, Chinese consumers are the biggest spenders in the luxury industry. In 2020, the global luxury market declined at 23%; however, China's share of the market nearly doubled, growing from about 11% in 2019 to 20% in 2020 (Bain & Company, 2020).

In 2019, the purchases of Chinese tourists in Italy increased by 15% in contrast to a loss of 7% during the previous year. This growth involved all Italian cities, but to a greater extent in Florence and Venice, while Milan remained the main one shopping destination. On average, Chinese tourists spent EUR1,217 in tax-free purchases per person in Italy, a much higher amount than the average spending of tourists from other countries, which was EUR862 (Fondazione Italia-Cina, 2020).

At the beginning of 2020, Italy and China planned to strengthen their partnership further by promoting the China-Italy Year of Culture and Tourism. However, the COVID-19 pandemic has disrupted such plan, so the celebration was postponed to 2022 (Travel No Stop, 2020). As pointed out by scholars recently (Wen et al., 2020), COVID-19 is expected to affect Chinese travelers' consumption patterns, pushing for free and independent travel, luxury trips, and health and wellness tourism. Therefore, once global tourism is resumed, Italian companies operating in the tourism industry and also the supporting local institutions should consider this evolution.

According to the Italy–China Foundation, the habits of Chinese tourists in Italy will go in the direction of seeking quality experiences. Indeed, in 2018, expenses for meals, hotels, and cultural and entertainment activities have grown, representing 14%, 21%, and 24% of total consumption, respectively (Fondazione Italia-Cina, 2020). These data confirm what emerged from studies conducted in the academic literature (Corigliano, 2011; Pearce et al., 2013) regarding the drop in spending on luxury and consumer goods, which decreased from 34% to 16%. Chinese tourism is therefore expected to evolve further. For Chinese travelers, local experiences, visits to museums, quality accommodation, and restaurants will be likely to become as relevant activities as tourism shopping, which traditionally characterized the Chinese tourism in Italy and Europe.

In this new context, also digital platforms will have a major role. Recently, several Italian institutions have developed their official account on social media like WeChat to provide information to Chinese tourists before booking the trip and during the travel experience (Hu, 2019). Indeed, in the traveler's customer journey, digital touchpoints are involved in all the phases: pre-purchase, purchase, and post-purchase (Hu & Olivieri, 2020). Online reviews contribute to generate electronic word-of-mouth (eWOM) in the travel industry (Filieri & McLeay), affecting travelers' destination choice. Therefore, tourism and travel firms should develop a strong digital strategy to reach and engage Chinese tourists.

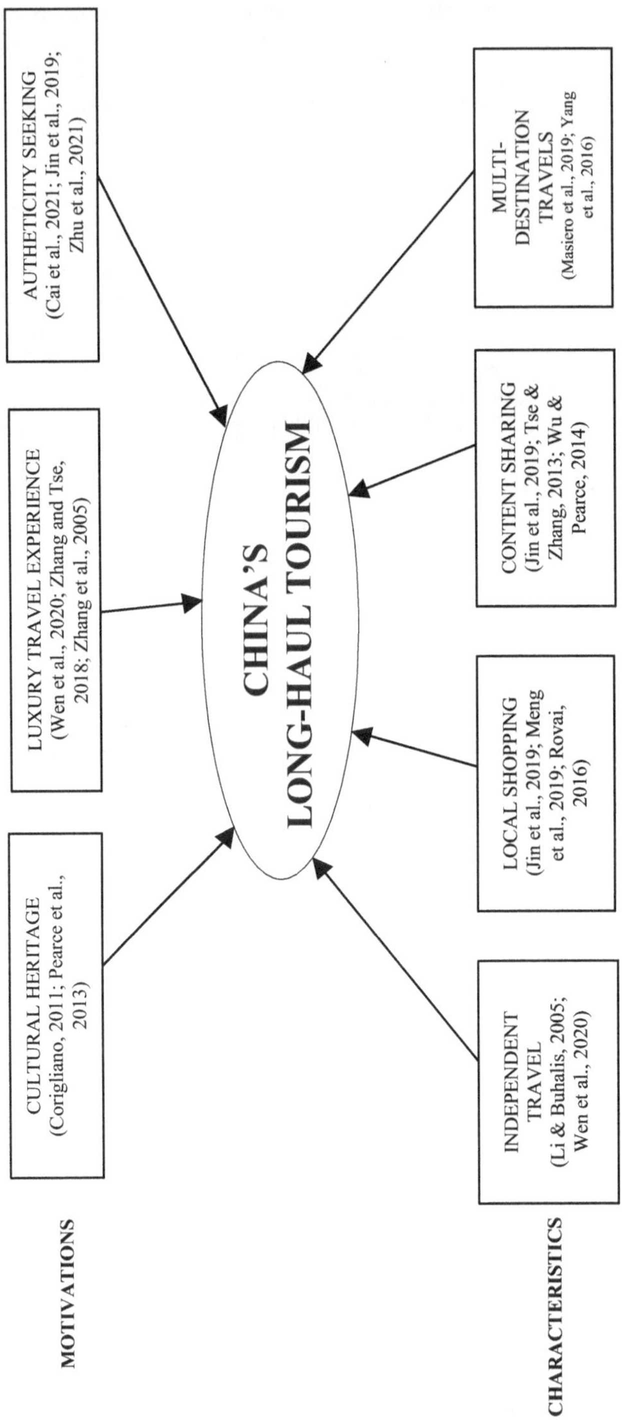

Figure 6.2 A conceptual framework of China's long-haul tourism

China's long-haul tourism: A conceptual framework

In Figure 6.2, we present a conceptual framework of China's long-haul tourism built on the revised literature and data on the travel evolution of Chinese tourists in Italy. In particular, we consider the following dimensions of the long-haul travel experiences of Chinese tourists: (1) Motivations and (2) Characteristics.

Regarding motivations, as discussed in the "Chinese tourists' characteristics" section, we identified three main ones:

1 *Cultural heritage.* Our analysis suggests that the Italian rich cultural heritage is a determining factor in the choice of Chinese long-haul travelers (Corigliano, 2011; Pearce et al., 2013). Most visited Italian cities are Venice, Florence, Rome, and Milan.
2 *Authenticity seeking.* In recent years, Chinese tourists tend to seek more authentic products and experiences that are typical of the local destination, for example, through dining tourism (Cai et al., 2021).
3 *Luxury travel experience.* For Chinese tourists, traveling abroad continues to symbolize a high level of social status and prestige (Zhang & Tse, 2018; Zhang et al., 2005). Therefore, the Chinese long-haul tourism will be more and more characterized by the search for unique experiences, also related to health and wellness (Wen et al., 2020).

Concerning the characteristics of the Chinese long-haul tourism, four main ones have been identified:

1 *Multi-destination travels.* The concept of multi-destination trips has been intensely analyzed by previous scholars (Lue et al., 1993; Masiero et al., 2019). With regard to Chinese tourists, they usually book multi-destination trips as there are few long holidays during the year (e.g., Chinese New Year and Golden Week). This allows to reach multiple destinations during the long-haul trip as well as to combine different types of experiences.
2 *Local shopping.* Shopping represents an important activity for Chinese outbound tourists since it is a hedonic and enjoyment setting (Meng et al., 2019) and allows them to access local authentic products (Jin et al., 2019). Purchasing foreign products in their COO can provide two main benefits. First, it can increase the perception of their exclusivity (Rovai, 2016). Second, price advantages are provided since foreign brands' prices in the COO are lower than how they would cost in China due to import duties.
3 *Content sharing.* In the tourism context, Chinese tourists heavily rely on online information while planning their travel. During and after their trips, travelers also like sharing information and tips on travel blogs (Jin et al., 2019; Tse & Zhang, 2013; Wu & Pearce, 2014).

4 *Independent travel.* The use of blogs and digital platforms has increased the popularity of independent travel in China (Li & Buhalis, 2006). This trend has been further amplified by COVID-19 effects since after the self-isolation of the lockdown period, Chinese tourists prefer to travel independently or in small groups thus reducing risks due to traveling in large groups (Wen et al., 2020). Moreover, independent travelers seem to be strongly driven by authenticity seeking than other tourism segments (Cai et al., 2021).

From this study, managerial implications for professionals working in this sector emerge. First, operators in the tourism industry should consider the specific characteristics of Chinese tourists highlighted in our conceptual framework to design their tourism offers and services. In particular, marketing and communication strategies of the tourism companies providing long-haul travels should design exclusive and personalized cultural experiences. Tourism firms should develop a digital presence on platforms used by Chinese tourists to book their travels and engage with them during their experiences. Indeed, social media and miniblogs are frequently used by Chinese tourists to share information and travel routes (Jin et al., 2019; Tse & Zhang, 2013; Wu & Pearce, 2014). Finally, destination organizations and marketers should also address tourists' concern over health risks due to COVID-19 by implementing measures to reduce travelers' overall perceived risk (Li et al., 2020).

Conclusion

The purpose of this theoretical study is both to attempt to advance the academic literature on tourism by providing a conceptual framework of China's long-haul tourism with a focus on Italy as a main destination and to yield managerial implications for companies operating in this sector.

In particular, our conceptual framework identifies three main dimensions of long-haul tourism, that is, cultural heritage, authenticity seeking, and luxury travel experience (Cai et al., 2021; Corigliano, 2011; Pearce et al., 2013; Wen et al., 2020; Zhang & Tse, 2018; Zhang et al., 2005). From our analysis, it emerged that Chinese travelers are driven by a strong interest in cultural heritage, arts, and history in the destination country. Moreover, they seek experiences that are typical of the local destination. Finally, long-haul travel market is more exclusive and upscale than domestic travel. Four main characteristics have emerged among Chinese tourists, that is, independent travel, travel shopping, content sharing, and multi-destination travels. All dimensions included in our conceptual framework highlight that the consumer behavior of Chinese on vacation in Italy will increasingly consider quality experiences. Together with the trends that have traditionally driven Chinese tourism in Italy, such as shopping, our study reveals new developments that these tourists appreciate – for example, local experiences, visits to museums, quality accommodation, and restaurants will become relevant activities (Fondazione Italia-Cina, 2020). In addition, the recent COVID-19 pandemic has significantly affected the consumption habits of Chinese tourists who had to replace long-haul travels with short-haul travel. This represents a challenge for tourism business operators once global tourism resumes.

In conclusion, with this study, we attempt to provide a critical perspective on the evolution of the Chinese long-haul tourism by reviewing the literature and systematizing the findings of the research into a conceptual framework. Some limitations are present due to the focus of our analysis and the type of data considered. As we only considered Chinese tourists' behavior in a single destination country (i.e., Italy), findings might not be generalizable. Future research could compare this analysis with the behavior of Chinese tourists in other long-haul destinations such as France. Moreover, we did not use primary data; therefore, the conceptual framework should be empirically tested by adopting both qualitative and quantitative methodologies. Finally, future studies could not only focus on consumers but also on the perspective of tourism companies operating to understand how they manage the growth of tourism coming from China.

References

Arlt, G. W. (2006). *China's outbound tourism*. Oxford: Routledge.

Bain & Company. (2020). *China's unstoppable 2020 luxury market*. Retrieved January 29, 2021, from https://www.bain.com/globalassets/noindex/2020/bain_report_chinas_unstoppable_2020_luxury-market.pdf

Bao, Y. F., & Mckercher, B. (2008). The effect of distance on tourism in Hong Kong: A comparison of short haul and long-haul visitors. *Asia Pacific Journal of Tourism Research, 13*(2), 101–111. doi:10.1080/10941660802048332

Baum, T., & Hai, N. T. T. (2020). Hospitality, tourism, human rights and the impact of COVID-19. *International Journal of Contemporary Hospitality Management, 32*(7), 2397–2407. doi:10.1108/ijchm-03-2020-0242

Bonetti, E., Petrillo, C. S., & Simoni, M. (2006). Tourism system dynamics: A multi-level destination approach. In L. Lazzaretti & C. S. Petrillo (Eds.), *Tourism local systems and networking* (pp. 111–131). Amsterdam: Elsevier.

Cai, W., Gebbels, M., & Wan-Zainal-Shukri, W. H. (2021). Performing authenticity: Independent Chinese travellers' tourism dining experiences in Europe. *Tourism Management, 86*, 104339. doi: 10.1016/j.tourman.2021.104339

China Internet Network Information Center. (2021). *Statistical report on Internet development in China*. Retrieved February 5, 2021, from https://cnnic.com.cn

Corigliano, M. A. (2011). The outbound Chinese tourism to Italy: The new graduates' generation. *Journal of China Tourism Research, 7*(4), 396–410. doi:10.1080/19388160.2011.627015

Dai, B., Jiang, Y., Yang, L., & Ma, Y. (2017). China's outbound tourism—Stages, policies and choices. *Tourism Management, 58*, 253-258. doi:10.1016/j.tourman.2016.03.009

Eurostat. (2018). *Tourism statistics - EU and China*. Retrieved January 29, 2021, from https://ec.europa.eu/eurostat/statistics-explained/index.php?title=Tourism_statistics_-_EU_and_China#Tourism_from_China_to_the_EU_has_tripled_in_ten_years

Filieri, R., & McLeay, F. (2014). E-WOM and accommodation: An analysis of the factors that influence travelers' adoption of information from online reviews. *Journal of Travel Research*, *53*(1), 44–57. doi:10.1177/0047287513481274

Fondazione Italia-Cina. (2020). *XI Rapporto Annuale CeSIF. Scenari e prospettive per le imprese*. Retrieved January 30, 2021, from https://www.fondazioneitaliacina.it/

Fry, R. B., & Prentice-Dunn, S. (2005). Effects of a psychosocial intervention on breast self-examination attitudes and behaviors. *Health Education Research*, *21*(2), 287–295. doi:10.1093/her/cyh066

Gagliardi, A. (2020, September 14). Nel 2020 turisti dimezzati, ritorno a livelli pre-COVID nel 2023. In vacanza sei italiani su dieci. *Il Sole 24 Ore*. Retrieved January 29, 2021, from https://www.ilsole24ore.com/art/nel-2020-turisti-calo-46percento-spesi-68-miliardi-meno-58percento-italiani-vacanza-autunno-ADn0TIp

Gu, Q., & Huang, S. S. (2018). Profiling Chinese wine tourists by wine tourism constraints: A comparison of Chinese Australians and long-haul Chinese tourists in Australia. *International Journal of Tourism Research*, *21*(2), 206–220. doi:10.1002/jtr.2255

Ho, G., & McKercher, B. (2014). A comparison of long-haul and short-haul business tourists of Hong Kong. *Asia Pacific Journal of Tourism Research*, *19*(3), 342–355. doi:10.1080/10941665.2012.746235

Hu, L. (2019, November 18). Perché chi fa turismo in Italia deve aprire un account Wechat. *Wired.it*. Retrieved January 29, 2021, from https://www.wired.it/economia/business/2019/11/18/wechat-turismo/#:~:text=Perch%C3%A9%20Wechat&text=Attraverso%20un%20mini%2Dprogram%20pensato, sportivi%20o%20dedicati%20allo%20shopping.

Hu, L., & Olivieri, M. (2020). Social media management in the traveller's customer journey: An analysis of the hospitality sector. *Current Issues in Tourism*, *24*(12), 1768–1779. doi:10.1080/13683500.2020.1819969

Istituto Nazionale di Statistica. (2020). *Report Turismo*. Retrieved January 29, 2021, from https://www.istat.it/it/files//2020/12/REPORT_TURISMO_2020.pdf

Jin, H., Moscardo, G., & Murphy, L. (2019). Exploring Chinese outbound tourist shopping: A social practice framework. *Journal of Travel Research*, *59*, 004728751982630. doi:10.1177/0047287519826303

Keating, B., & Kriz, A. (2008). Outbound tourism from China: Literature review and research agenda. *Journal of Hospitality and Tourism Management*, *15*(1), 32–41. doi:10.1375/jhtm.15.32

Lew, A. A., & McKercher, B. (2002). Trip destinations, gateways and itineraries: The example of Hong Kong. *Tourism Management*, *23*(6), 609–621. doi:10.1016/s0261-5177(02)00026-2

Li, C. (Spring), McCabe, S., & Li, X. (Robert). (2017). Digging deeper into decision-making of Chinese long-haul outbound tourists: A two-stage preference-estimation approach. *Journal of Destination Marketing & Management*, *6*(3), 267–275. doi:10.1016/j.jdmm.2016.09.004

Li, L., & Buhalis, D. (2006). E-commerce in China: The case of travel. *International Journal of Information Management*, *26*(2), 153–166. doi:10.1016/j.ijinfomgt.2005.11.007

Li, X. (Robert), Meng, F., Uysal, M., & Mihalik, B. (2013). Understanding China's long-haul outbound travel market: An overlapped segmentation approach. *Journal of Business Research*, *66*(6), 786–793. doi:10.1016/j.jbusres.2011.09.019

Li, Z., Zhang, S., Liu, X., Kozak, M., & Wen, J. (2020). Seeing the invisible hand: Underlying effects of COVID-19 on tourists' behavioral patterns. *Journal of Destination Marketing & Management*, *18*, 100502. doi:10.1016/j.jdmm.2020.100502

Lin, V. S., Liu, A., & Song, H. (2015). Modeling and forecasting Chinese outbound tourism: An econometric approach. *Journal of Travel & Tourism Marketing*, *32*(1–2), 34–49. doi:10.1080/10548408.2014.986011

Lo, A., & Lam, T. (2004). Long-haul and short-haul outbound all-inclusive package tours. *Asia Pacific Journal of Tourism Research*, *9*(2), 161–176. doi:10.1080/1094166042000233694

Lue, C. C., Crompton, J. L., & Fesenmaier, D. R. (1993). Conceptualization of multi-destination pleasure trips. *Annals of Tourism Research*, *20*(2), 289–301.

Masiero, L., Qiu, R. T. R., & Zoltan, J. (2019). Long-haul tourist preferences for stopover destination visits. *Journal of Travel Research*, *59*, 1–17. doi:10.1177/0047287519867143

McKercher, B. (2008). Segment transformation in urban tourism. *Tourism Management*, *29*(6), 1215–1225. doi:10.1016/j.tourman.2008.03.005

McKinsey. (2018, September). *Chinese tourists: Dispelling the myths.* Retrieved January 29, 2021, from https://www.mckinsey.com/~/media/McKinsey/Industries/Travel%20Transport%20and%20 Logistics/Our%20Insights/Huanying%20to%20the%20new%20Chinese%20traveler/Chinese-tourists-Dispelling-the-myths.ashx

Meng, F., Zhang, P., Li, H., & So, K. K. F. (2019). Modeling precursors of impulsive tourist shopping behavior: Evidence from long-haul Chinese outbound tourists. *International Journal of Tourism Research*, *21*, 1–15. doi:10.1002/jtr.2266

Paris, C. M. (2012). Flashpackers: An emerging sub-culture? *Annals of Tourism Research*, *39*(2), 1094–1115. doi:10.1016/j.annals.2011.12.001

Pearce, P. L., Wu, M.-Y., & Chen, T. (2015). The spectacular and the mundane: Chinese tourists' online representations of an iconic landscape journey. *Journal of Destination Marketing & Management*, *4*(1), 24–35. doi:10.1016/j.jdmm.2014.11.001

Pearce, P. L., Wu, M.-Y., De Carlo, M., & Rossi, A. (2013). Contemporary experiences of Chinese tourists in Italy: An on-site analysis in Milan. *Tourism Management Perspectives*, *7*, 34–37. doi:10.1016/j.tmp.2013.04.001

Prayag, G., Cohen, S. A., & Yan, H. (2015). Potential Chinese travellers to Western Europe: segmenting motivations and service expectations. *Current Issues in Tourism*, *18*(8), 725–743. doi:10.1080/13683500.2013.868413

PricewaterhouseCoopers. (2021). *L'itinerario del turismo tra ripartenza ed evoluzione.* Retrieved January 29, 2021, from https://www.pwc.com/it/it/about-us/Italia2021/doc/Italia2021-IlTurismo-itinerario-del-turismo-tra-ripartenza-ed-evoluzione.pdf

Rovai, S. (2016). Luxury tourism: The new shopping perspective. In Serena Rovai (Ed.), *Luxury the Chinese way: New competitive scenarios* (pp. 113–122). London: Palgrave Macmillan.

Skift Research. (2020, September). *The travel industry turned upside down: Insights, analysis and actions for travel executives.* Retrieved January 29, 2021, from https://www.mckinsey.com/~/media/mckinsey/industries/travel%20transport%20and%20logistics/our%20insights/the%20travel%20industry%20turned%20upside%20down%20insights%20analysis%20and%20actions%20for%20 travel%20executives/the-travel-industry-turned-upside-down-insights-analysis-and-actions-for-travel-executives.pdf

Sparks, B., & Pan, G. W. (2009). Chinese outbound tourists: Understanding their attitudes, constraints and use of information sources. *Tourism Management*, *30*(4), 483–494. doi:10.1016/j.tourman.2008.10.014

Statista. (2020a). *Travel and tourism industry in China*. Retrieved January 29, 2021, from https://www.statista.com/study/12813/travel-and-tourism-market-in-china-statista-dossier/

Statista. (2020b). *Number of overnight stays by tourists from China (including Hong Kong) in travel accommodation in Italy from 2008 to 2019*. Retrieved January 29, 2021, from https://www.statista-com.opac.unicatt.it/statistics/901296/number-of-nights-spent-by-chinese-tourists-in-accommodations-in-italy/

Statista. (2020c). *Arrivals of tourists from China (including Hong Kong) in tourist accommodation in Italy from 2008 to 2018*. Retrieved January 29, 2021, from https://www.statista.com/statistics/901197/number-of-arrivals-from-china-in-tourist-accommodations-in-italy/

Travel No Stop. (2020, December 3). *Italia e Cina dialogano su Anno Cultura e Turismo 2022*. Retrieved January 29, 2021, from https://travelnostop.com/news/territori/italia-e-cina-dialogano-su-anno-cultura-e-turismo-del-2022_507149

Tse, T. S. M., & Hobson, J. S. P. (2008). The forces shaping China's outbound tourism. *Journal of China Tourism Research*, *4*(2), 136–155. doi:10.1080/19388160802279459

Tse, T. S. M., & Zhang, E. Y. (2013). Analysis of blogs and microblogs: A case study of Chinese bloggers sharing their Hong Kong travel experiences. *Asia Pacific Journal of Tourism Research*, *18*(4), 314–329. doi:10.1080/10941665.2012.658413

Tussyadiah, I. P., Kono, T., & Morisugi, H. (2006). A model of multidestination travel: Implications for marketing strategies. *Journal of Travel Research*, *44*(4), 407–417. doi:10.1177/0047287505282950

United Nations Educational, Scientific and Cultural Organization. (2021). *Number of world heritage properties by region*. Retrieved October 29, 2021, from http://whc.unesco.org/en/list/stat

United Nations World Tourism Organization. (2020). *International tourism highlights*. Retrieved February 6, 2021, from https://www.e-unwto.org/doi/epdf/10.18111/9789284422456

We Are Social. (2020). *Digital 2020 China*. Retrieved February 20, 2021, from https://wearesocial-cn.s3.cn-north-1.amazonaws.com.cn/digital2020-china.pdf

Wen, J., Kozak, M., Yang, S., & Liu, F. (2020). COVID-19: Potential effects on Chinese citizens' lifestyle and travel. *Tourism Review*, *76*(1), 74–87. doi:10.1108/tr-03-2020-0110

Wu, M.-Y., & Pearce, P. L. (2014). Tourism blogging motivations. Why do Chinese tourists create little "Lonely Planets"? *Journal of Travel Research*, *55*(4), 537–549. doi:10.1177/0047287514553057

Xu, Y., & McGehee, N. G. (2012). Shopping behavior of Chinese tourists visiting the United States: Letting the shoppers do the talking. *Tourism Management*, *33*(2), 427–430. doi:10.1016/j.tourman.2011.05.003

Yang, C.-H., Lin, H.-L., & Han, C.-C. (2010). Analysis of international tourist arrivals in China: The role of World Heritage Sites. *Tourism Management*, *31*(6), 827–837. doi:10.1016/j.tourman.2009.08.008

Yang, Y., Fik, T. J., & Zhang, H. (2016). Designing a tourism spillover index based on multidestination travel: A two-stage distance-based modeling approach. *Journal of Travel Research*, *56*(3), 317–333. doi:10.1177/0047287516641782
Zhang, E. Y., & Tse, T. S. M. (2018). Tapping into Chinese Luxury Travelers. *Journal of China Tourism Research*, *14*(1), 71–99. doi:10.1080/19388160.2018.1437102
Zhang, H. Q., Pine, R., & Lam, T. (2005). *Tourism and hotel development in China: From political to economic success*. Oxford: The Haworth Hospitality Press.
Zhu, J. S., Airey, D., & Siriphon, A. (2021). Chinese outbound tourists as international consumer in Northern Thailand—A dynamic mobility perspective. *Journal of Consumer Culture*, 1469540521994318. doi:10.1177/146954052199431.

7 Experience economy in wellness tourism to attract Chinese outbound tourists

Yedan Fan, Salmi Mohd Isa and Shaohua Yang

Introduction

Due to the COVID-19 pandemic and restrict measures of international travel, the wellness tourism in Malaysia is in its stagnation and retrogression of development. Moreover, since the COVID-19 was reported first in China and then spread worldwide, the potential influences of the pandemic were not only limited to health and safety of human being but also included psychological and behavioral changes among Chinese outbound tourists. Since the Chinese outbound tourists had been the third largest inbound market for Malaysia, it's quite desirable for the destination marketing organizations in Malaysia, especially the wellness tourism sector, to develop a new strategy to attract the Chinese outbound tourists to respond to pandemic environment. Therefore, in this chapter, economic impacts of COVID-19 pandemic on wellness tourism in Malaysia are analyzed, and potential influences of COVID-19 on Chinese outbound tourists' attitude and behavior are identified. Further, four realms of experience economy are introduced on the basis of the natural advantages of wellness tourism in Malaysia, which is an implication for development of Malaysia tourism during post-COVID-19 pandemic.

Chinese inbound tourism market in Malaysia

The Malaysia Tourism Promotion Board (MTPB), known as Tourism Malaysia (TM), is established to promote tourism marketing activities domestically and internationally (TM, 2020a). According to their report, "Key Performance Indicators 2019", China ranked third in the top ten markets list with 3.11 million inbound tourists' arrivals and contributed RM 15.3 billion to Malaysian GDP (TM, 2020a). In recent years, Malaysia has witnessed a rapidly increasing amount of Chinese inbound tourists from 1,130,261 in 2010 to 3,114,257 in 2019, which has nearly tripled during that period (TM, 2020a). Almost 83.5% of the Chinese inbound tourists chose to travel in Malaysia for their vacations, while another 11.9% part of them enjoyed shopping during the trip (TM, 2020a).

DOI: 10.4324/9781003121329-9

Based on the official report, "Tourism Profile", provided by TM, Chinese outbound tourists selected Malaysia as their destination due to their wants and needs to learn more about Malaysia as well as famous tourism attractions, along with shopping environment (TM, 2020b). Overall, Malaysia is one of the favorable tourism destinations to Chinese outbound tourists, and the Chinese outbound tourism market will continually be an important segment in terms of its volume and expenditure for Malaysia.

The economic influence of COVID-19 pandemic on Malaysia's wellness tourism

The outbreak of the COVID-19 had become a global crisis in terms of economic loss and health emergency. During the whole year of 2020, Malaysian government had imposed four phases of Movement Control Order (MCO) and Recovery Movement Control Order (RMCO) as an effective measure to prevent the spread of COVID-19 (Khan & Hashim, 2020). But the pandemic in Malaysia has not been controlled well; thus, in 2021, a more rigorous MCO 3.0 is imposed beginning from June 1 to June 14 and then extended to June 21, 2021 including a period of 14 days of total lockdown nationwide (FMCO) (Ong, 2021). Statistics from TM have shown a terrible economic loss in the tourism sector.

First, the overall tourist arrivals in 2020 in Malaysia declined to 4.33 million with a decreased growth rate of −83.4% (TM, 2021a). Further, the total expenditure decreased to 12.7 billion (RM) with a growth rate of −85.3% compared to the previous year (TM, 2021a). In addition, the tourist receipts reduced to 12.7 billion (RM), indicating a negative growth rate of −85.2% (TM, 2021a). Specifically, in 2020, only 405,149 Chinese outbound tourists chose Malaysia as the tourism destination, followed by −86.99% of growth rate (TM, 2021b). The whole tourism industry suffered from a great deal of economic loss. As a sector of the tourism industry, the wellness tourism was inevitably influenced by the pandemic. The wellness tourism failed to achieve its goal which aims to support the Visit Malaysia 2020 campaign to bring in 30 million international tourists and RM 100 billion in tourist receipts to the country (TM, 2021c). The wellness tourism, as well as other sector of the tourism industry, is in its stagnation and retrogression of development due to the significant impact of COVID-19.

The potential impacts of COVID-19 on Chinese outbound tourists

Given the fact that the COVID-19 has led to a global crisis, the impacts of COVID-19 extend from tourism to tourists in terms of image, psychology and consumer behavior (Richburg, 2020; Wen et al., 2020; Zheng et al., 2020). The COVID-19 pandemic was first reported in China, which was labeled as "Chinese virus pandemonium" by biased media (World Health

Organization [WHO], 2020; Yang et al., 2021; Zheng et al., 2020). The biased media reports had portrayed a negative image of China as well as the Chinese, whereas Chinese outbound tourists were perceived as threats of health to overseas since the virus and the pandemic were viewed as associated with the Chinese (Armutlu et al., 2021; Richburg, 2020; Yang et al., 2021). Furthermore, Zheng et al. (2020) demonstrated that biased media reports about COVID-19 misled the public to believe that the Chinese should be avoided, put Chinese outbound tourists in the ethically center of perceived "social isolation" with unequal status and would negatively influence Chinese outbound tourists' mental well-being to some extent.

In regard to other threatened potential impact of the long-lasting pandemic, "perceived risk" and "uncertainty" are identified, which influence Chinese outbound tourists' attitude and behavior toward their choices of destinations (Ghosh, 2020; Wassler & Talarico, 2021; Yu et al., 2021; Zheng et al., 2020). Conceptually, "perceived risk" is highly associated with "uncertainty" since perceived risk is derived from uncertain and unpredictable consequences, especially unpleasant and negative outcomes (Shin & Kang, 2020). "Perceived risk" is defined as "the subjective evaluation of undesirable loss or consequences result from the uncertainty of a product or service" (Shin & Kang, 2020; Yu et al., 2021). As tourism is an "intangible" and "experiential" service, absence of risk during the travel experience has been acknowledged to accelerate positive emotions and subjective well-being for those pleasure-seeking tourists (Williams & Baláž, 2015). Adversely, high level of uncertainty and perceived risk are critical factors that distort tourists' decision-making processes and behaviors (Shin & Kang, 2020; Yu et al., 2021).

In addition, the pandemic has potentially reshaped Chinese outbound tourists' attitudes and behavioral patterns of traveling. Tourists may increase their awareness of health on food safety, healthy diets, enjoyment of outdoors and a return to "spirituality" to reflect on human being and self-concept (Zwanka & Buff, 2021). During the post-pandemic era, health and well-being-related tourism would be a popular choice for Chinese outbound tourists with a higher level of healthy awareness (Wen et al., 2020). For example, in the short term, Chinese outbound tourists may be more likely preferential to independent travel or small group with relatives, families or friends to keep social distance during their trips. In the long term, outdoor activities which enhance body health, wellness tourism which extends travel with more physical and mental health experiences would be more attractive to Chinese outbound tourists (Wen et al., 2020).

Rational tourists tend to maximize the utilitarian value of consumption and select uncertainty avoidance behavior. Particularly, in the period of the post-COVID-19 pandemic, Chinese outbound tourists are more sensitive of perceived physical health risks and psychological risks as well as uncertainties in tourism overseas (Wen et al., 2020; Yu et al., 2021). Therefore, Chinese tourists are more cautious to traveling outbound to avoid negative image perceived by countries overseas as well as to reduce the risks of traveling. Besides, aligning with the improvement of health awareness, wellness tourism or health-related tourism products may be preferred by Chinese outbound tourists to counteract risks and uncertainties during post-COVID-19 pandemic. To further facilitate the recovery of tourism during the post-pandemic period, destination marketing organizations should make deliberate efforts to eliminate potential risks and uncertainties in tourism, especially for the Chinese outbound tourists, who are used to be perceived as threats and be blamed to be responsible for the pandemic.

Natural advantages of wellness tourism in Malaysia

"Wellness tourism" is an ancient form of traveling dated from 18th to 19th century when European elite seeking for "spiritual enlightenment" in terms of spa tourism or medical trip (Smith & Kelly, 2006). Defined by Global Wellness Institute, "wellness tourism" means traveling connected with "the pursuit of maintaining or enhancing one's personal well-being" (Global Wellness Institute [GWI], 2021). Wellness tourism is distinctive from another form of health tourism: medical tourism in terms of the healthy state of a person. In medical tourism, people who are suffering from disease are motivated to seek a cost-effective treatment through traveling, whereas in wellness tourism, people actively decide to travel for the purpose of enhancement of wellness and pursuit of a better state of well-being (Dillette et al., 2021; Hashemi et al., 2015; Mueller & Kaufmann, 2001).

To some extent, "wellness tourism" is a journey rather than a destination since a tourist is an "active health seeker" who is highly motivated to self-enhancement and self-recovery during the trip (Smith & Kelly, 2006). Although wellness tourism is still a unique type of lifestyle to relax and escape from daily routine and release negative stress or depression, it's more than that, and it has been valued as a format of self-awareness and self-fulfillment (Han et al., 2020; Smith & Kelly, 2006). Voigt et al. (2010) identified six dimensions of benefits of wellness tourism: transcendence, physical health and appearance, escape and relaxation, important others and novelty, re-establishing self-esteem and indulgence. In wellness tourism, tourists not only experience the hedonic value but also the eudaimonic value as to experience good life (Vada et al., 2019; Voigt et al., 2010; Winter et al., 2019). Experiences in wellness tourism have been proven that have a long-lasting effect on tourists after gaining the hedonic and eudaimonic values (Smith & Kelly, 2006; Voigt et al., 2010).

As a sub-stream of health tourism, wellness tourism is predominantly prevailing in Asia, especially in Malaysia, etc., which is increasingly developed depending on abundant natural resources and unique healthy cultures (Han et al., 2020; Hashemi et al., 2015). Geographic advantages with prolific marine and tropical forest resources, historical and cultural heritages, unique Muslim religion in Malaysia make the country itself charming and favorable

(Cartier, 1998). Abundant nature-based tourism resources, ranging from marine scenery, beaches to jungles, waterways, wildlife, etc., are the backbone of booming and development since interactions with natural physical environment can elicit the sense of well-being, which is beneficial to tourists physically and mentally (Mueller & Kaufmann, 2001; Nair & Thomas, 2013; TM, 2020a, 2020b). Traditional therapies, unique Muslim culture and extraordinary climate with fascinating natural environments are packaged as a wellness tourism product provided by resorts or nature-based destinations in Malaysia (Hashemi et al., 2015; TM, 2020a, 2020b).

Adapt the theory of experience economy into wellness tourism in Malaysia

Unique, happy, memorable experience is viewed as a venue of competitive advantages (Hosany & Witham, 2010; Pine & Gilmore, 1998). It's particularly important for wellness tourism sector to follow the experience-dominant logic to stage experiences in the lens of experience economy to achieve well-being and loyalty (Ali et al., 2016; Hwang & Lee, 2019; Hwang & Lyu, 2015). "Experience-dominant logic" believes that experience is the key element of business since every tourist experience is different in terms of subjective feelings and sequential behavioral intentions; thus, designing personalized experience is considered as an experience-center logic (Gentile et al., 2007; Pine & Gilmore, 1998; Schmitt, 1999). The theory of experience economy is a typical representative of "experience-dominant logic". Based on the level of participation (active vs. passive) and on the level of connection with environment or activities/events (absorption or immersion), Pine and Gilmore (1998) identified four realms of experience economy consisting of entertainment, education, esthetic and escapism experience.

Entertainment experience includes events that can stimulate and encourage tourists passively participate in as tourists take the form of observers or listeners, so they can be affected with enjoyment and delightful sensations (Kim et al., 2019; Pine & Gilmore, 1998). Tourists' participation does not change or influence the external entertainment activities or events; thus, they are absorbed rather than immersed (Kim et al., 2019; Oh et al., 2007). Education experience is related to the needs and wants of human natural desire to learn something new or acquire some skills (Hosany & Witham, 2010; Oh et al., 2007). Education experience relates to the needs and wants of human natural desire to learn something new or acquire some skills (Hosany & Witham, 2010; Oh et al., 2007). Education experience is a form of active and absorptive participation as new knowledge or new skills appeal tourists' internal desire to attend with more enthusiasm, and education experience can "occupy a person's attention by bringing the experience into the mind" (Hosany & Witham, 2010; Pine & Gilmore, 1999, p. 31).

Esthetic experience reflects the immersion in an activity or external environment, in which tourists passively appreciate the natural or artificial art of environment or events while they make no impact during the experience (Kim et al., 2019; Oh et al., 2007; Pine & Gilmore, 1998). The esthetic experience highlights the importance of the influence of external environment on tourists' physical and mental well-being during a wellness trip (Hwang, & Lyu, 2015; Kim et al., 2019). Escapism experience is a kind of completely immersion and actively involvement in the environment and activities, which means that tourists are not only appealed into the fascinating external environment but also act an important role in the activity or interact with the environment (Kim et al., 2019; Luo et al., 2018). In escapism experience, tourists feel a sense of relief, be away from their daily routine life, participate in a new lifestyle even for a short period of time and fulfill "self-center" needs to some extent (Hosany & Witham, 2010; Hwang & Lyu, 2015; Luo et al., 2018).

The four realms of experience are mutually compatible and combined to form uniquely individual experience, but the four dimensions differ in terms of relative importance in explaining consumer attitudes and behavioral intentions (Hosany & Witham, 2010). Since well-being is a consequence of an experience (Lyu et al., 2018; Vada et al., 2019), to satisfy tourists' needs of well-being in wellness tourism, the first and foremost thing is to stage diversified experiences in the lens of experience economy to create memorable experience for tourists, whereas tourists with memorable experience are more likely to achieve a higher level of loyalty (Ali et al., 2014, 2016; Kim et al., 2019).

The wellness tourism is naturally advantageous to develop experience economy. Abundant nature-based tourism resources have laid foundations on staging experience economy since esthetic experience and escapism experience highlight the absorption or immersion in environment (Hosany & Witham, 2010; Pine & Gilmore, 1998). Moreover, the wellness tourism includes activities such as spa or sunbathing, which can extend to more well-being-related activities to develop entertainment experience and education experience to improve the level of enjoyment and fun and to enrich new knowledge or skills for tourists (Ali et al., 2016; Pine & Gilmore, 1998). From the perspective of marketing wellness tourism, well-being is inevitably associated with experiences, whereas the theory of experience economy provides a strategy of staging experiences with different realms. Staging entertainment experience, education experience, esthetic experience and escapism experience in wellness tourism is beneficial to create unique memories and satisfy the needs of well-being of tourists (Ali et al., 2016; Hosany & Witham, 2010; Oh et al., 2007).

Value of experience economy in wellness tourism

As consumers are seeking fun, fantasy and enjoyment in the consumption experience (Holbrook & Hirschman, 1982), the experience economy can engage consumers at "an emotional, physical, intellectual or even spiritual level"; therefore, experience staged in the lens of the experience economy is evolved as a new business competitive advantage (Pine & Gilmore, 1998; Sundbo & Sørensen, 2013). The tourism industry, due to the intangible and experiential nature, has been at the heart of staging experiences (Gilmore & Pine, 2002; Oh et al., 2007). The value of experience economy has been proven in a plethora of studies (Andersson, 2007; Hosany & Witham, 2010; Oh et al., 2007), which provides a great explanation on positive emotion as well as consumer satisfaction, vivid memories, positive emotion, well-being perception, as well as loyalty, including positive word-of-mouth and revisit behavior or behavioral intentions (Ali et al., 2016; Chang, 2018; Kim et al., 2019; Loureiro, 2014; Sidali et al., 2015; Song et al., 2015).

The four realms of experience economy are naturally adaptable in wellness tourism, whereas the experiences can contribute to well-being as well as loyalty (Hwang & Lee, 2019; Hwang & Lyu, 2015; Luo et al., 2018). Ali et al. (2016) suggested to develop the experience economy in resort hotel context as the four realms of experience economy could significantly promote positive emotions of tourists which, in turn, contributed to a high level of satisfaction with the experiences. Kim et al. (2019) found that the four dimensions of experience economy, especially the entertainment experience, could significantly influence brand prestige and ultimately lead to loyalty. Hwang and Lee (2019) explained that distinct realms of experience contributed to well-being differently, whereas their findings suggested that the most significant factor which affected well-being perception was education experience, followed by entertainment experience, esthetic experience and escapism experience among senior tourists. The experience economy is naturally adaptable to the wellness tourism; in the current situation, marketing experiences associated with well-being can help to reduce risks and uncertainties, to promote confidence of tourists and to rise positive emotions and feelings (Ali et al., 2016; Hwang & Lyu, 2015; Song et al., 2015).

Wellness tourism marketing among Chinese outbound tourists

The Global Wellness Institute (GWI, 2021) defines wellness as "the active pursuit of activities, choices and lifestyles that lead to a state of holistic health". Wellness is a balanced state of health which means the harmony of body, mind, spirit and environment (Dillette et al., 2021; Mueller & Kaufmann, 2001). GWI expands the "holistic wellness" into six dimensions: physical, mental, emotional, spiritual, social and environmental, which enriches its connotation (GWI, 2021). Chinese outbound tourists are also familiar with the concept of "wellness", while a similar concept "*Yangsheng*" is prevailing among the Chinese. "*Yangsheng*" represents an ancient form of traditional culture together with Confucianism (Huang & Xu, 2014). The cultural beliefs of "*Yangsheng*" are rooted in the Chinese culture and convention over 5,000 years. It can be achieved through a balanced state of "body-mind holism" and "body-mind-nature holism" (Huang & Xu, 2014).

Similarly, "*Yangsheng*" advocates the harmony of internal self with external environment, which is an art of self-cultivation (Huang & Xu, 2014; Ren et al., 2007). "*Yangsheng*" facilitates self-healing, self-growth with both physical and mental health as well as a positive state of well-being (Huang & Xu, 2014; Ren et al., 2007).

Compared with people in other countries and regions, the pursuit of health and well-being of the Chinese people has been rooted and practiced with the development of Chinese history, and it seems that products and services in wellness tourism would be more popular among Chinese outbound tourists, in particular, in the post-pandemic era (Han et al., 2020; Huang & Xu, 2014). Overall, for those Chinese outbound tourists, the wellness tourism in Malaysia could be an attractive destination in the post-pandemic era, whereas marketing a dramatic strategy of experience economy associated with well-being could be effective to stimulate the desire of wellness tourism among Chinese outbound tourists (Vada et al., 2019; Voigt et al., 2011).

Conclusion

As China is the third largest inbound tourist market in Malaysia, to recover from the pandemic, the Chinese outbound tourist market is a necessary competitive marketplace. Destination marketing organizations in Malaysia should first understand the potential impacts of COVID-19 on Chinese outbound tourists, including afraid of perceived negative image in overseas, afraid of perceived risks and uncertainties, which lead to a high level of health awareness and preference of health-related tourism (Wen et al., 2020; Yu et al., 2020). Tourism sectors in Malaysia need to pay attention to these potential effects and make efforts to reduce and eliminate these uncertainties, potential risks to facilitate the recovery and booming development of the tourism industry.

For the wellness tourism, as the Chinese outbound tourists are familiar and adapt to the concept of "wellness", in addition, the Chinese outbound tourists are more likely to prefer health-related tourism products, and it implicates a new strategy to marketing experiences with well-being in wellness tourism among Chinese outbound tourists. The wellness tourism sector in Malaysia should conduct appropriate strategies in marketing wellness tourism on the basis of natural resources, for example, highlight the concept of nature, health, well-being or "yangsheng" among Chinese outbound tourists (Huang & Xu, 2014; Ren et al., 2007).

Furthermore, well-being results from experiences, to ensure the long-last competitive advantages and to achieve long-term success, not only just recover from the pandemic but also the wellness tourism in Malaysia should develop with a memorable experience-based orientation (Ali et al., 2016; Pine & Gilmore, 1998). An important strategy is to adapt the experience economy into the wellness tourism in Malaysia, which is effective to exert memorable experiences for tourists as well as satisfying the needs of well-being (Ali et al., 2016; Lyu et al., 2018; Vada et al., 2019). Entertainment experience, education experience, esthetic experience and escapism experience are developed based on intentionally staged or themed activities, events and performances as well as environmental elements. The wellness tourism is experiential in nature and developed on the basis of natural resources and unique culture of Muslin in Malaysia. Destination marketing organizations should take account to the preconditions first and deliberately design different realms of experience economy as a strategy to develop wellness tourism. Wellness tourism marketers should build a match between tourists' psychological needs of well-being, involving physical health and mental health and the staged experience economy to create desired and memorable experiences for Chinese outbound tourists.

At the national level, tourism policymakers in Malaysia should consider the wellness sector as well as other sectors in terms of recovery and development. Tourism policymakers should first make efforts to eliminate the risks and uncertainties that reside in international traveling to ensure a safe environment for Chinese outbound tourists. Moreover, facilitating measures of international travel should be prepared before opening the inbound tourism market. In addition, the wellness tourism should be supported and encouraged as a representative sector of recovery with more economic measures such as tax reduce or tax preference. In terms of the individual organizations in the wellness sector, experience economy should be well studied with organizational resources and conditions. Each organization should carefully integrate, coordinate and allocate its resources appropriately to stage different dimensions of experience economy and to exert the strategy of the experience economy successfully.

This chapter only discussed the recovery strategy of wellness tourism from the perspective of economy, whereas other strategies from the view of political, cultural and environmental should also be vital for tourism recovery in Malaysia. Further studies should address the limitations from other perspectives, e.g., arranging the effective measures to ensure safety in traveling as well as controlling the spread of the virus. As this chapter is based on the discussion of prior studies and the current issues of COVID-19 pandemic, international tourism is still limited due to the restriction measures, and it is suggested that further studies could test the experience economy in wellness tourism among Chinese outbound tourists to examine its effect on well-being, which can extend the knowledge of experience economy with empirical evidences in Malaysia.

References

Ali, F., Hussain, K., & Omar, R. (2016). Diagnosing customers experience, emotions and satisfaction in Malaysian resort hotels. *European Journal of Tourism Research*, *12*, 25–40. https://ejtr.vumk.eu/index.php/about/article/view/211

Ali, F., Hussain, K., & Ragavan, N. A. (2014). Memorable customer experience: Examining the effects of customers experience on memories and loyalty in Malaysian resort hotels. *Social and Behavioral Sciences, 144*, 273–279. doi:10.1016/j.sbspro.2014.07.296

Andersson, T. D. (2007). The tourist in the experience economy. *Scandinavian Journal of Hospitality and Tourism, 7*(1), 46–58. doi:10.1080/15022250701224035

Armutlu, M. E., Bakır, A. C., Sönmez, H., Zorer, E., & Alvarez, M. D. (2021). Factors affecting intended hospitable behaviour to tourists: Hosting Chinese tourists in a post-COVID-19 world. *Anatolia: An International Journal of Tourism and Hospitality Research, 32*(2), 218–231. doi:10.1080/13032917.2020.1855595

Cartier, C. (1998). Megadevelopment in Malaysia: From heritage landscapes to "leisurescapes" in Melaka's tourism sector. *Singapore Journal of Tropical Geography, 19*(2), 151–176. doi:10.1111/j.1467–9493.1998.tb00257.x

Chang, S. (2018). Experience economy in hospitality and tourism: Gain and loss values for service and experience. *Tourism Management, 64*, 55–63. doi:10.1016/j.tourman.2017.08.004.

Dillette, A. K., Douglas, A. C., & Andrzejewski, C. (2021). Dimensions of holistic wellness as a result of international wellness tourism experiences. *Current Issues in Tourism, 24*(6), 794–810. doi:10.1080/13683500.2020.1746247

Gentile, C., Spiller, N., & Noci, G. (2007). How to sustain the customer experience: An overview of experience components that co-create value with the customer. *European Management Journal, 25*(5), 395–410. doi:10.1016/j.emj.2007.08.005.

Ghosh, S. (2020). Asymmetric impact of COVID-19 induced uncertainty on inbound Chinese tourists in Australia: Insights from nonlinear ARDL model. *Quantitative Finance and Economics, 4*(2), 343–364. doi:10.3934/QFE.2020016

Gilmore, J. H., & Pine, B. J. (2002). Differentiating hospitality operations via experiences: Why selling services is not enough. *The Cornell Hotel and Restaurant Administration Quarterly, 43*(3), 87–96. doi:10.1177/0010880402433009

Global Wellness Institute. (2021). Wellness definitions. Retrieved September 5, 2021, from https://globalwellnessinstitute.org/what-is-wellness/wellness-economy-definitions/

Han, H., Kiatkawsin, K., Koo, B., & Kim, W. (2020). Thai wellness tourism and quality: Comparison between Chinese and American visitors' behaviors. *Asia Pacific Journal of Tourism Research, 25*(4), 424–440. doi:10.1080/10941665.2020.1737551

Hashemi, S. M., Jusoh, J., Kiumarsi, S., & Mohammadi, S. (2015). Influence factors of spa and wellness tourism on revisit intention: The mediating role of international tourist motivation and tourist satisfaction. *International Journal of Research- GRANTHAALAYAH, 3*(7), 1–11. doi:10.29121/granthaalayah.v3.i7.2015.2976

Holbrook, M. B., & Hirschman, E. C. (1982). The experiential aspects of consumption: Consumer fantasy, feelings and fun. *Journal of Consumer Research, 9*(2), 132–140. doi:10.1086/208906

Hosany, S., & Witham, M. (2010). Dimensions of cruisers' experiences, satisfaction, and intention to recommend. *Journal of Travel Research, 49*(3), 351–364. doi:10.1177/0047287509346859

Huang, L., & Xu, H. (2014). A cultural perspective of health and wellness tourism in China. *Journal of China Tourism Research, 10*(4), 493–510. doi:10.1080/19388160.2014.95175

Hwang, J., & Lee, J. (2019). A strategy for enhancing senior tourists' well-being perception: Focusing on the experience economy. *Journal of Travel & Tourism Marketing, 36*(3), 314–329. doi:10.1080/10548408.2018.1541776

Hwang, J., & Lyu, S. O. (2015). The antecedents and consequences of well-being perception: An application of the experience economy to golf tournament tourists. *Journal of Destination Marketing & Management, 4*, 248–257. doi:10.1016/j.jdmm.2015.09.002

Khan, M. A. A., & Hashim, H. (2020). The effect of COVID-19 on tourism and hospitality industry in Malaysia, resurgence in the post-pandemic era: A conceptual criterion. *International Journal of Tourism & Hospitality Review, 7*(2), 54–62. doi:10.18510/ijthr.2020.726

Kim, S., Ham, S., Moon, H., Chua, B. L., & Han, H. (2019). Experience, brand prestige, perceived value (functional, hedonic, social, and financial), and loyalty among GROCERANT customers. *International Journal of Hospitality Management, 77*, 169–177. doi:10.1016/j.ijhm.2018.06.026

Loureiro, S. M. C. (2014). The role of the rural tourism experience economy in place attachment and behavioral intentions. *International Journal of Hospitality Management, 40*, 1–9. doi:10.1016/j.ijhm.2014.02.010.

Luo, Y., Lanlung, C., Kim, E., Tang, L. R., & Song, S. M. (2018). Towards quality of life: The effects of the wellness tourism experience. *Journal of Travel & Tourism Marketing, 35*(4), 410–424. doi:10.1080/10548408.2017.1358236

Lyu, J., Mao, Z., & Hu, L. (2018). Cruise experience and its contribution to subjective well-being: A case of Chinese tourists. *International Journal of Tourism Research, 20*(2), 225–235. doi:10.1002/jtr.2175

Mueller, H., & Kaufmann, E. L. (2001). Wellness tourism: Market analysis of a special health tourism segment and implications for the hotel industry. *Journal of Vacation Marketing, 7*(1), 5–17. doi:10.1177/135676670100700101

Nair, P. K., & Thomas, T. K. (2013). Sustainable tourism in Malaysia: Policies and practices. *Mondes du Tourisme, 8*, 59–69. doi:10.4000/tourisme.88

Oh, H., Fiore, A. M., & Jeoung, M. (2007). Measuring experience economy concepts: Tourism applications. *Journal of Travel Research, 46*(2), 119–132. doi:10.1177/0047287507304039

Ong, E. (2021). Complete list of official nationwide MCO SOP 2021. *Klook*. Retrieved October 15, 2021, from https://www.klook.com/en-MY/blog/malaysia-pkpp/

Pine, B. J., & Gilmore, J. H. (1998, July–August). Welcome to the experience economy. *Harvard Business Review*.

Pine, B. J., & Gilmore, J. H. (1999). *The experience economy: Work is theatre & every business a stage*. Boston, MA: Harvard Business School Press.

Ren, H., Xing, W. H., & Fu, F. H. (2007). The concept of wellness from the perspective of Chinese culture. *Journal of Exercise Science and Fitness, 5*(2), 95–101.

Richburg, K. B. (2020). China coronavirus. Is history repeating itself? *Statnews*. Retrieved August 27, 2020, from https://www.statnews.com/2020/01/27/bird-flu-sars-china-coronavirus-is-history-repeating-itself/

Schmitt, B. (1999). Experiential marketing. *Journal of Marketing Management, 15*(1–3), 53–67. doi:10.1362/026725799784870496

Shin, H., & Kang, J. (2020). Reducing perceived health risk to attract hotel customers in the COVID-19 pandemic era: Focused on technology innovation for social distancing and cleanliness. *International Journal of Hospitality Management, 91*, 102664. doi:10.1016/j.ijhm.2020.102664

Sidali, K. L., Kastenholz, E., & Bianchi, R. (2015). Food tourism, niche markets and products in rural tourism: Combining the intimacy model and the experience economy as a rural development strategy. *Journal of Sustainable Tourism*, *23*(8/9), 1179–1197. doi:10.1080/09669582.2013.836210

Smith, K., & Kelly, C. (2006). Wellness tourism. *Tourism Recreation Research*, *31*(1), 1–4. doi:10.1080/02508281.2006.11081241

Song, H. J., Lee, C. K., Park, J. A., Hwang, Y. H., & Reisinger, Y. (2015). The influence of tourist experience on perceived value and satisfaction with temple stays: The experience economy theory. *Journal of Travel & Tourism Marketing*, *32*(4), 401–415. doi:10.1080/10548408.2014.898606

Sundbo, J., & Sørensen, F. (2013). Introduction to the experience economy. In J. Sundbo & F. Sørensen (Eds.), *Handbook on the experience economy* (pp. 1–18). Cheltenham: Edward Elgar. doi:10.4337/9781781004227.00005

Tourism Malaysia. (2020a). *Key performance indicators 2019*. Retrieved August 10, 2021, from https://www.tourism.gov.my/files/uploads/KPI2019-web-nov2020.pdf

Tourism Malaysia. (2020b). *Tourism profile*. Retrieved August 10, 2021, from https://www.tourism.gov.my

Tourism Malaysia. (2021a). *Malaysia tourism statistics in brief*. Retrieved August 10, 2021, from https://www.tourism.gov.my/statistics

Tourism Malaysia. (2021b). *Arrivals by country*. Retrieved August 12, 2021, from http://mytourismdata.tourism.gov.my/?page_id=232#!range=year&from=2019&to=2020&type=55876201563fe, 558762c48155c&destination=3&origin=32CN, 34SG

Tourism Malaysia. (2021c). *MATTA and AMSPA committed to promote wellness tourism in Malaysia*. Retrieved August 12, 2021, from https://www.tourism.gov.my/news/trade/view/matta-and-amspa-committed-to-promote-wellness-tourism-in-malaysia

Vada, S., Prentice, C., & Hsiao, A. (2019). The influence of tourism experience and well-being on place attachment. *Journal of Retailing and Consumer Services*, *47*, 322–330. doi10.1016/j.jretconser.2018.12.007

Voigt, C., Brown, G., & Howat, G. (2011). Wellness tourists: In search of transformation. *Tourism Review*, *66*(1/2), 16–30. doi: 10.1108/16605371111127206

Voigt, C., Howat, G., & Brown, G. (2010). Hedonic and eudaimonic experiences among wellness tourists: An exploratory enquiry. *Annals of Leisure Research*, *13*(3), 541–562. doi:10.1080/11745398.2010.9686862

Wassler, P., & Talarico, C. (2021). Sociocultural impacts of COVID-19: A social representations perspective. *Tourism Management Perspectives*, *38*, 100813. doi:10.1016/j.tmp.2021.100813.

Wen, J., Kozak, M., Yang, S., & Liu, F. (2020). COVID-19: Potential effects on Chinese citizens' lifestyle and travel. *Tourism Review*, *76*(1), 74–87. doi:10.1108/TR-03-2020-0110

Williams, A. M., & Baláž, V. (2015). Tourism risk and uncertainty: Theoretical reflections. *Journal of Travel Research*, *54*(3), 271–287. doi:10.1177/0047287514523334

Winter, P. L., Selin, S., Cerveny, L., & Bricker, K. (2020). Outdoor recreation, nature-based tourism, and sustainability. *Sustainability*, *12*, 81. doi:10.3390/su12010081

World Health Organization. (2020). *Coronavirus disease (COVID-19) outbreak situation view dashboard*. Retrieved August 23, 2021, from https://www.who.int/emergencies/diseases/novel-coronavirus-2019

Yang, S., Isa, S. M., & Ramayah, T. (2021). How are destination image and travel intention influenced by misleading media coverage? Consequences of COVID-19 outbreak in China. *Vision: The Journal of Business Perspective*, 1–10. doi:10.1177/0972262921993245

Yu, J., Lee, K., & Hyun, S. S. (2021). Understanding the influence of the perceived risk of the coronavirus disease (COVID-19) on the post-traumatic stress disorder and revisit intention of hotel guests. *Journal of Hospitality and Tourism Management*, *46*, 327–335. doi:10.1016/j.jhtm.2021.01.010

Zheng, Y., Goh, E., & Wen, J. (2020). The effects of misleading media reports about COVID-19 on Chinese tourists' mental health: A perspective article. *Anatolia: An International Journal of Tourism and Hospitality Research*, *31*(2), 337–340. doi:10.1080/13032917.2020.1747208

Zwanka, R. J., & Buff, C. (2021). COVID-19 generation: A conceptual framework of the consumer behavioral shifts to be caused by the COVID-19 pandemic. *Journal of International Consumer Marketing*, *33*(1), 58–67. doi:10.1080/08961530.2020.1771646

8 The challenges of targeting the Chinese outbound market

A case study of Fiji

Sera Vada and Noel Scott

Introduction

Tourism is an important driver of the Fijian economy, providing more than 118,500 jobs and contributing approximately 34% towards Fiji's gross domestic product in 2019 (Ministry for Industry, Trade & Tourism, 2020). Tourism in Fiji has also been a key driver of the economy despite the political coups in 1987, 2000, and 2006 (Harrison & Pratt, 2010). However, since the development of tourism in Fiji in the 1920s, the main source markets for Fiji's tourism industry have been Australia, New Zealand, and the United States (U.S.). Dependence on these traditional source markets (Hall & Page, 1996) is a concern as any economic recession or unease in the political environment in these traditional markets may likely result in the decline of international tourist arrivals to Fiji.

The China outbound travel market has been one of the fastest growing international markets over the past decade (Jang et al., 2003; Li et al., 2011), and according to the United Nations World Tourism Organization (UNWTO), China will be the world's top tourist destination by 2020, receiving a total of 140 million international visitors annually. In addition, China is expected to rank as the fourth largest outbound tourist generating country contributing 100 million visitors annually to the world travel market. The increase in Approved Destination Status (ADS) destinations has also widened the range of choices for Chinese travellers (Xie & Li, 2009). Chinese travel expenditure is also expected to be the second fastest growing in the world at close to twice the global average (UNWTO, 2011).

Many tourism destinations have resorted to the diversification of source markets to increase their tourist arrivals and reduce risk (Agrusa et al., 2011). Recently, a growing number of tourism destinations have turned towards the Chinese outbound market as a potential source market to offset the risk of a decline in visitor arrivals from traditional source markets that have been affected by exogenous shocks (Song & Lin, 2009) such as the Global Financial Crisis. The option to expand into the Chinese outbound market is considered attractive since China's economy is less likely to be affected by global financial shocks relative to other countries due to its closed financial system.

DOI: 10.4324/9781003121329-10

Much of the tourism literature has focused on tourists' perceptions, and level of satisfaction when travelling in Western countries and few studies specifically focus on Asian countries. Over the past decades, there have been a growing number of studies carried out to identify and understand the motivations of Chinese outbound tourists and their decision-making process to attract this growing outbound market (Hanqin & Lam 1999; Kim et al., 2012; Lo & Lam, 2004; Pan & Laws, 2007). Whilst much research has been carried out on the implications of the Chinese outbound market in developed countries such as Australia (Pan & Laws, 2007; Sparks & Pan, 2009), New Zealand (Zhao, 2006), and the U.S., there has been limited research carried out on the Chinese outbound tourists to Pacific Island Countries (PICs). For Fiji, the decision to focus on the Chinese outbound market has been attributed to the rapid growth in visitor arrivals from China. There have been limited evident quantitative or qualitative studies conducted on the Chinese outbound market to Fiji. This lack of research carried out on the Chinese outbound market especially to PICs and Fiji is addressed in this chapter.

The Chinese tourists have a very different cultural background compared to Western tourists and therefore have different preferences and motivations (Zhang & Heung, 2002). Whilst there have been existing studies, whereby cultural models of tourist behaviour have been constructed to explain their behaviour (Reisinger & Turner, 2002; Weiermair, 2000), there have been limited quantitative or qualitative studies carried out on the Chinese inbound market to Fiji and, in particular, on analysing and understanding the characteristics of Chinese tourists travel motivations and visitor experiences in Fiji. In addition, there is also no evident research of the perceptions of key tourism stakeholders in Fiji on the growing number of Chinese tourists to Fiji, and this information is critical to determine whether it is worthwhile for Fiji to tap into the China outbound market and whether Fiji has the appropriate facilities and suitable services to cater to the Chinese market. It is clearly stated in Fiji's Tourism Development Plan 2017–2016 that "there is potential for new direct links with China, but the potential demand of Fiji's tourism product needs to be assessed" (Ministry of Tourism, 2007, p. 40).

Literature review

Tourism has become a solution to less-developed countries and small island developing states alike due to its potential to promote economic growth, increase employment opportunities, and improve social conditions. For many PICs, including Fiji, tourism has progressed to the forefront of economic growth and development (Harrison, 2003; Rao, 2002; Reddy, 2010). In particular, Fiji's tourism industry is the largest of all South Pacific nations, and on average, Fiji receives more than double the number of tourist arrivals per annum than its nearest competitor, French Polynesia (South Pacific Tourism Organisation, 2009).

Tourism began during Fiji's colonial days when steamship routes through Fiji provided the impetus for local Europeans to establish tourism for consumers (Hall & Page, 1996). It is important to note that Fiji's tourism industry relies heavily on the cooperation of the local populations (Kanemasu, 2013). In its infancy stage, tourism in Fiji is primarily capitalized on its tropical climate and natural environment. However, from the late 1960s onwards, the reputation of the Fiji Islanders as naturally hospitable and friendly people created a key competitive advantage of Fiji's tourism product (Samy, 1980).

Fiji was one of the first Pacific nations to embrace tourism after World War II, and its long history and experience with tourism are characterized by a relatively strong and effective national tourism organization. Currently, Fiji has a much wider set of source markets than its major competitors with most of the Fiji's tourist arrivals coming from Australia, New Zealand, the U.S., Europe, Japan, Korea, and China. In comparison, Fiji's main competitors, namely Maldives, Mauritius, and French Polynesia, are largely dependent on long-haul markets in Europe and South Africa (Narayan, 2000). Given its geographic proximity, Australia and New Zealand continue to be the Fiji's dominant source markets with a combined market share of 70%, whilst Asia is a growing market for Fiji with China maintaining its market share of 4% in 2013.

Diversification to the Chinese outbound market

Many destinations are targeting Chinese tourists using aggressive marketing and promotion (Tse & Hobson, 2008). For example, within Asia, China has already become the largest source market of outbound tourism and has already become New Zealand's fourth and Australia's fifth largest source market (Li et al., 2011). Chinese tourist numbers to the South Pacific region are small with only Fiji noting a significant rise in arrivals over the last decade. In 2009, Fiji recorded 4,087 visitor arrivals from China (0.8% of the total visitors), and in 2012, 26,395 (Fiji Bureau of Statistics, 2013).

In 2009, the Chinese market was one of the Fiji's strongest performers and is expected to grow in the future (Fiji Times, 2009). The Counsellor of the Embassy of the Republic of China reaffirmed that the new direct flight from Nadi to Hong Kong was a positive move in helping to increase the number of tourists from China. In addition, Fiji was the essence of a tropical island paradise with white sandy beaches, swaying coconut trees, pristine ocean and waterways, and a range of things to do and see that would appeal to the most discerning traveller (Serelini, 2010).

Likewise, South Pacific Tourism Organization Chief Executive Officer Ilisoni Vuidreketi stated that part of the South Pacific region's plan is to attract some of the 400 million visitors that are expected to visit the East Asia Pacific region in the next ten years, which accounts for 25% of the 1.6 billion visitors that will travel the world. He also mentioned that the National Airline's direct flights to Hong Kong and Korean Airlines' non-stop service to Seoul made marketing Fiji as a destination easier for the Chinese outbound market (Baselala, 2010). China has been identified as an emerging market for Fiji. The friendship between the Fijian and Chinese people dates back to before Fiji was ceded to Great Britain in 1874 when the first Chinese migrants made Fiji their home. In 1975, Fiji became the first Pacific Island Country to establish diplomatic relations with the People's Republic of China which was followed by the signing of the joint communique in 1976 when the Chinese Government established its Embassy in Suva.

Fiji first entered the Chinese outbound market in 2004 with the signing of the Approved Destination Status (ADS) between the Government of the Republic of Fiji and the Government of the People's Republic of China. However, it was not until 2007 that visa exemptions for Chinese nationals visiting Fiji were established. According to Opperman and Chon (1997), visa regulations can hinder tourism development, sometimes purposely. Relaxation or removal of visa requirements is a facilitator of tourism development.

In 2009, the National Airline, Fiji Airways (former Air Pacific), provided an additional service to Hong Kong to tap into the larger Chinese market. This was one of the objectives highlighted in the Fiji Tourism Development Plan (TDP) 2007–2016, where it is mentioned that an air link with Hong Kong or a mainland China port is essential (Ministry of Tourism, 2007). In addition, in 2010, Fiji adopted an increasingly proactive foreign policy 'Look North – One China Policy' to strengthen its relations with Asia, in general, and China, in particular. Therefore, this growing recognition of the potential of the Chinese inbound market to Fiji has prompted more tourism operators in Fiji to tailor their products and services to suit this market by hiring Chinese coordinators to assist with language interpretation and bookings, providing signage in the Chinese language as well as extending office hours in line with the time difference in China to allow for flexibility in bookings.

In recent years, however, with more and more Chinese tourists travelling abroad, some tourism destinations have experienced confronting Chinese tourists' behaviour that has caused concern and criticism (Gong et al., 2018; Zhang, 2013). For example, Chinese tourists defacing an ancient temple in Egypt; throwing cigarette boxes into the sea in the Republic of Palau; soaking feet in the fountain outside the Louvre Museum in Paris; and Chinese graffiti on London's St Paul's Cathedral (Zhang, 2013).

In response to these recent negative reports of Chinese tourists' poor behaviour, in 2013, the Chinese National Tourism Administration issued a 64-page etiquette handbook that includes tips on what is acceptable and unacceptable behaviour during an international tour. In addition, China Central Television also aired educational clips aimed at promoting polite tourist behaviour (Xinhua, 2007). Research has suggested that etiquette is heavily dependent on culture (Fraser, 1990), and what is excellent etiquette in one society may be a shock in another.

The importance of tourist stakeholders

Destination management cannot be successful without the involvement and effective coordination of key stakeholders. There have been several tourism studies that have highlighted the importance of tourist stakeholder cooperation (Scott et al., 2008; Vernon et al., 2005). Garcia et al. (2012) suggest that stakeholders are divided into primary and secondary stakeholders. Primary stakeholders have direct interaction and impact on the brand equity, whilst secondary stakeholders have an indirect interaction. Morrison (2013) suggests that there are five main groups of stakeholders in destination management: tourists, tourism sector organizations, community, environment, and Government. Understanding the working relationships among tourism businesses in a destination is a critical prerequisite to the success of many collaborative destination-marketing programmes (Wang & Krakover, 2008). On one hand, the tourist destination offers a unified tourist product about other destinations; on the other hand, within the destination, there is competition between the different elements of the tourist product. Therefore, the success and development of the destination can be affected by whether the different stakeholders collaborate.

Destination marketing and the promotion of the tourism destination in key target markets are common but a critical activity that all tourism stakeholders undertake. Destination branding also plays a broader role and includes the network of stakeholders within a destination. According to Sautter and Leisen (1999), DMOs should not only consider those stakeholders consistent with the guidelines of the strategic plans of tourism but also other perspectives or services in a wider scope that are indirectly associated with tourism. In addition, the sustainability of the destination brand is related not only to the message delivered but also to a degree of shared meaning that is contained in the message delivered through the brand. Therefore, all stakeholders within a tourism destination need to have a shared contribution to upholding a destination's brand and communicating this effectively to its target markets (Foley & Fahy, 2004).

However, despite the rising interest in stakeholders, effective stakeholder involvement is complex, problematic, and often underestimated (Friedman & Miles, 2006). This is mostly due to the existence of multiple and diverse stakeholders that often hold different viewpoints. Furthermore, Waligo et al. (2013) stated that it is also imperative to recognize stakeholders when managing tourism more sustainably and to take account of their different perspectives on the issues. More specifically, the DMO needs to know what processes work because tourism development can result in the heavy exploitation of a locality's resources by the developers, the visitors, and other users if not managed well.

Challenges of tourism development in Fiji

There are fundamental challenges to maintaining the competitive position of Fiji in providing a perceived safe destination, being price competitive, and remaining attractive for investors. Three critical challenges are highlighted. They are the uneven distribution of tourism earnings and development throughout the country, the absence of certain destination attributes which appeal specifically to the Chinese outbound market, and the heavy dependence of Fiji on its traditional markets.

First, although tourism is the top earner of foreign exchange in Fiji, it has not reached its potential to contribute effectively to poverty reduction and development (Native Land Trust Board, 2006). More specifically, whilst the tourism industry in Fiji has been successful in generating substantial amounts of foreign investment, the development has been uneven, and benefits have not necessarily trickled down to the poorest groups. For example, Fiji's 2017 TDP concentrated the tourism industry by focusing on major tourist regions in Fiji. As a result, a recommendation from the Strategic Environmental Assessment of Fiji's TDP was that as a matter of national priority, all development activities should spread economic benefits to all parts of the country with efforts to promote tourism.

For example, tourist development is more aligned to small-scale community-based ecotourism, run or jointly owned by local populations as this type of tourism is more reliant on local expertise as well as products (World Wide Fund, 2003). It is also suggested that greater benefits from tourism can be spread to the poor by encouraging a wider range of players (community, private sector, civil society, government) working at a range of scales (local, national, regional) to spread the benefits of tourism more widely and unlock livelihood opportunities for the poor within the tourism and connected sectors. This would ultimately lead to improvements in policy and labour practices of hotels and resorts.

Second, research carried out by Stollznow Research on the competitive position of Fiji showed that compared to other locations, Fiji does not perform well on destination attributes such as 'shopping', 'lots to see and do', and 'nightlife'. For Chinese tourists, in particular, it is noted that shopping is the top priority for many Chinese tourists (Boycoff, 2014), and Chinese travel expenditure is expected to be the second fastest growing in the world at close to twice the global average. As clearly noted in the Fiji TDP 2007–2016, creating a demand for Fiji from China will be a challenge (Ministry of Tourism, 2007), and major efforts to include cuisine and shopping products to match demand are essential.

Recent reports highlight that one of the major reasons for the Chinese to travel abroad is to buy luxury products. This is due to the high tariffs that China charges for imported luxury products. However, shopping in Fiji is not found in modern malls and high-end department stores with luxury-branded products. Instead, this activity is common within resorts, hotels, and boutiques along city streets and is mostly limited to handicraft products such as hand-woven baskets, mats, and jewellery. Recent reports (China International Travel Monitor, 2014), however, have found that although Chinese tourists do more shopping than other tourists, it has also been found that scenery and culture are becoming bigger draw cards that influence Chinese tourists to visit a particular tourism destination. Therefore, although shopping in Fiji is quite limited compared to other developed countries, the destination does possess certain attributes that appeal to Chinese tourists more than shopping.

The heavy dependence on the tourism industry, and in particular, on traditional markets of Australia, New Zealand, and the U.S. is also a cause for concern because the impacts of exogenous shocks in tourist source markets affect international travel. This was echoed in a statement by the Indonesia Tour and Travel Agencies Association, Bali chapter chairman, Aloysius Purwa, that "over-reliance on a single major market was similar to putting all of one's eggs in a basket. If something bad happens to the basket, you might end up with no eggs at all" (The Jakarta Post, 2011). Therefore, diversification has become a keyword for tourism destinations to ensure the resilience and sustainability of their tourism industry.

Finally, the challenges to tourism development in Fiji also depend on the collaboration between different stakeholders. Fiji as a tourist destination offers a unified tourist product about other destinations; however, within the destination, there is competition between the different elements of the tourist product. This is where destination management plays a critical role in facilitating a coordinated approach between key tourism stakeholders in Fiji.

It is important to note that it is stated in Fiji TDP 2007–2016 that the growth of tourism in Fiji and the target of 1.1 million visitors by 2016 do not appear to be possible without significant growth in new markets such as China, India, and South America (Ministry of Tourism, 2007). It is mentioned that the growth in tourist arrivals from these emerging markets relies heavily on creating the necessary air links and ensuring that Fiji offers tourism products that match demand, and this will require major branding and tactical marketing to create Brand Fiji. There is also the need to have a strong and coordinated collaboration between the National Airline, national tourist office, and major tourism products to develop these markets as well as new alliances with wholesalers and outbound operators. Specifically, for China, it is noted that there is a need for major efforts to create tourism products that meet Chinese demands, and these include cuisine and shopping products (Ministry of Tourism, 2007).

Methodology

This study conducted semi-structured face-to-face interviews from November 2013 to August 2014 with six key tourist stakeholders in Fiji that target the Chinese outbound market. These stakeholders were Tourism Fiji, Mana Island Resort, Rosie Holidays, Namale Island Resort, Musket Cove Island Resort, and Intercontinental Fiji Resort & Spa. Semi-structured interviews obtain information directly from the respondent based on an interview guide (Bernard, 2001). It is effective in studies with limited time and financial constraints (Patton, 2002). The interviews examined the perceptions of tourism stakeholders on whether it is worthwhile for Fiji to tap into the China outbound market and whether Fiji has the appropriate facilities and suitable services to cater to the Chinese market. The interviews also established feedback on the major challenges that are faced in meeting the demands and expectations of Chinese tourists to Fiji.

Results

The results from the interviews with the six key tourist stakeholders show that despite the positive perceptions about the growth and potential of the Chinese outbound market for Fiji, there are challenges with regards to marketing Fiji as a tourism destination and managing the demands and meeting expectations of Chinese tourists to Fiji. One of the major challenges highlighted when dealing with the Chinese market involved the high expectations of Chinese tourists:

> *Sometimes the guests are fussy and there are some problems. They complain. For example, the main thing is that they pay a lot of money to the agents and they come here with high expectations. And they only stay here in island bures with ocean views. It is quite different in China. In China, you pay a lot of money and you enjoy lots. So, the complaint is mostly about the island bures.*
>
> [personal communication]

The above complaint is justified by recent findings whereby hotels remained the most popular type of accommodation amongst Chinese tourists. More importantly, a hotel's branding and reputation, comfort, and facilities mainly drive the booking decision, whereas price is less of an issue (China International Travel Monitor, 2014). Likewise, other challenges identified were also based on the high expectation that Chinese tourists have of Fiji before they arrive:

> *I think a lot of it is all based on perception. They would like to travel to Fiji but they need to be aware of the varied activities that are available and also understand that although we are a small destination, there is quite a lot of logistics involved. So, I think that once they come down and do a familiarization trip, there is a lot more understanding on Fiji, what is involved, and they see the need for actually using a local operator to see their customers through their entire process.*
>
> [personal communication]

> *Compared to traditional markets where tourists know what they wanted to do, they went out surfing, they went out to swim. Whereas the Chinese markets, their wants are different, they don't want to go into the water, they want to go up to the mountains and they expect someone to take them.*

> *I think it is the expectation when they come to Fiji, from their perspective they pay so much and they expect so much. We don't know what has been promised to them, but when they come to this destination, we are promising them the world but in other traditional markets, we are promising them the world but they come knowing that the sun, sand, and sea are enough.*
>
> [personal communication]

However, there have been efforts to overcome this challenge or obstacle as services are being tailored according to the expectation of the Chinese inbound market to Fiji:

> *I guess the only challenge that we have is because we don't have a Chinese-speaking representative, language is an obstacle, but there is a huge amount of Chinese upcoming young generation who now speak English. Research tells us that we need to tailor content to suit your target audience and, in this case, it has to be in the Chinese language. So, our collaterals, our website, Nat Geo that they can read about Fiji, etc in the language that they best understand.*
>
> [personal communication]

Tourism stakeholders in Fiji are also beginning to recognize the need to employ Chinese coordinators to service this growing outbound market. For example, one stakeholder has two Chinese coordinators, and their responsibilities include checking in and checking out of Chinese guests, bookings for activities, and translation. Likewise, another key stakeholder has a Chinese department with 14 staff based locally in Fiji and 3 full-time staff based in Beijing. Some of the services they provide to Chinese tourists to Fiji include language translation, tour guiding services, meet and greet facilities, transfer, tours, and accommodation. They also cater to niche markets where there are Chinese guides that specifically look after this for the weddings, honeymoon, and dive markets.

Objectives for targeting the Chinese outbound market

Overall, there is a mixed reaction and feedback from key tourist stakeholders as there are different agendas and objectives for targeting the Chinese outbound market.

> *The tourism body should be the main organization driving the regulation of the market. However, Fiji does not have the people who can drive meetings with stakeholders on how to regulate this market. The only way to regulate this market is through the resorts and hotels or the properties on the ground.*
>
> [personal communication]

> *Destination marketing and management needs to be driven by the industry itself, and not just from the National Tourist Office directive, to follow a global trend and for Fiji to do the same.*
>
> [personal communication]

For the Ministry of Tourism and Tourism, Fiji, the main objectives are increasing the tourist numbers and maximizing the earnings. However, there is minimal concern highlighted in Fiji's National TDP 2007–2016 or marketing plans that address the evident problems with targeting the mass numbers of Chinese tourists and the associated economic, social, and environmental impacts that could arise as a result of inadequate planning. More importantly, there is no mention of the need for any preparatory work to be carried out beforehand in terms of educating the front-line service staff as well as the Chinese tourists themselves on what to expect during their trip to Fiji.

Understanding Chinese culture

A major challenge highlighted in interviews with key tourist stakeholders is the ability of tourism operators and especially front-line service providers in Fiji to first understand the Chinese culture and why Chinese tourists act the way they do. China has a very different cultural background than the Western countries, and this means that we should not take their preferences and motivations for granted and certainly should not expect theirs to be the same as the ones motivating people from the Western countries or even those from other parts of Asia (Zhang & Heung, 2002). Research has suggested that etiquette is heavily dependent on culture (Zhang, 2013), and what is excellent etiquette in one society may be a shock in another. With 5,000 years of civilization, China is supposed to have a culture that values courtesy and propriety (Zhou et al., 1998). The feedback from the two accommodation providers highlighted that the major difficulty in serving the Chinese market was educating their staff to deal with the different expectations and demands from Chinese tourists.

> *We realized that the Chinese market was quite a different market compared to traditional markets so we had to get our team to realize that this was a different market with different demands. They are not just numbers; this market is quite different with different expectations.*
>
> [personal communication]

> *Few things need to happen before Fiji targets this Chinese market. Fiji is ready for this market, but one of the things that we need to do is to understand the culture – why Chinese tourists do the things they do when they are in Fiji and for us not to be quick to pass judgment. When people in Fiji begin to understand the culture, looking after the Chinese culture would be much easier.*
>
> [personal communication]

The Chinese market is a different market with different expectations compared to traditional markets of Australia, New Zealand, and the U.S. Fiji has been targeting these traditional markets since the 1960s, and tourism operators and properties are familiar with the demands and expectations of these tourists especially since tourists from traditional markets are primarily repeating visitors.

> *What we realized is that the Chinese market is very loud, so what we did was coordinate separate dining arrangements. However, it was no expenses spared for Chinese tourists, as it was only a matter of service.*
>
> [personal communication]

> *When Chinese tourists came, they would be separated from tourists from traditional markets, and during cocktails, the staff would then announce to guests from traditional markets that Chinese tourists were staying at the property.*
>
> [personal communication]

According to one tourist stakeholder, tourists from traditional markets visit Fiji and know what they want to do in terms of activities in comparison to Chinese tourists. In addition, the tourists from traditional markets are satisfied with the sun, sea, and sand, whereas Chinese tourists have different expectations when they visit Fiji. If Fiji wants to attract Chinese travellers, the marketing will need to be done differently from that currently being done in countries like Australia and New Zealand, where sandy beaches and sunshine can attract tourists (Lal, 2011). Furthermore, there had been similar incidences experienced by two stakeholders, whereby Chinese tourists are loud, and thus, separate dining arrangements were made as well as announcements to tourists from traditional countries that there were Chinese tourists in-house as well.

Regulating the Chinese market to Fiji

The feedback from interviews with key tourist stakeholders is that there needs to be a more coordinated approach to targeting the Chinese outbound market and also in terms of regulating this emerging market so as not to encounter the same problems that other tourism destinations such as the Maldives have experienced in terms of mass numbers of Chinese tourists which have impacted other guests as well as creating social impacts on the host destination such as gambling. Tourism Australia unveiled its China 2020 Strategic Plan which aimed at nearly doubling the number of Chinese visitors to 860,000 over the next eight years. However, Australia is not promoting the cultural side of the destination, but instead promoting itself as a high-end destination for the super-rich, where Chinese tourists can stay in five-star hotels, buy luxury bags, and gamble in glitzy casinos (Ho, 2011). Some of the social impacts related to gambling include financial, personal, and family costs including financial hardships, debts, asset loss, bankruptcy, stress, depression, anxiety, suicide, poor health, the neglect of family, relationship breakdown, and violence (Deery et al., 2012).

> *Fiji has to learn from what we have been through in the past, with the Korean market and Japanese market and tour operators have learned from these. If we say yes to the Chinese market, then yes only to a particular time and that is during the low season. We know that the challenge is there but it is easier to deal with that challenge when we don't have a big number of traditional guests.*
>
> [personal communication]

> *We need to regulate this market to limit the numbers. Australia, Mauritius, Seychelles have all had an overflow of Chinese tourists. Fiji as a tourism destination needs to understand the lessons learned from other countries where other countries have not regulated this market and are now having problems.*
>
> [personal communication]

Some of the current strategies to regulate the Chinese inbound market to Fiji include opening up the resort to Chinese guests only during the low season. This ensures that the property capitalizes on revenue during the low season as well as having the resources, time, and ability to efficiently accommodate the different demands and expectations of the Chinese market. Likewise, another tourist accommodation provider regulates the influx of Chinese tourists by limiting the number of rooms available for Chinese guests, including separating Chinese guests from traditional guests during dining.

Discussion

A coordinated approach is needed between key tourist stakeholders as well as other ministries such as the Ministry of Transport and Ministry of Health for Fiji's tourism industry to be sustainable. DMOs should not only consider those stakeholders consistent with the guidelines of the strategic plans of tourism but also other services in a wider scope that are indirectly associated with tourism (Sautter & Leisen, 1999). In addition, gaining a share of the global tourism market is a complex and expensive process, thus requiring a significant degree of Government support (Sharpley, 2002), and although tourism has been generally regarded as a private sector activity, many governments should be active participants in tourism development (Fyall & Leask, 2006; Heath & Wall, 1992; Sofield, 2003)

Marketing activities for the Chinese outbound market

The funding and scale of marketing activities carried out in traditional markets of Australia and New Zealand are much more extensive in comparison to emerging markets such as China. This is because Fiji has been marketing itself as a tourism destination to Australia and New Zealand since the 1960s and that these source markets represent the top-ranking source markets for Fiji in terms of visitor arrivals and tourist earnings. It is important to also note that Tourism Fiji's regional offices are based in these traditional source markets and essentially drive the core marketing and awareness campaigns. This is in comparison to China which does not have a regional office but only a regional representative to work with key tourism stakeholders within the target cities of Beijing, Shanghai, and Guangzhou and spearhead all marketing activities for the Chinese outbound market.

China represents one of the most important emerging markets for Fiji, alongside India and Russia. However, it was evident that there were similar marketing activities that recur every year which focused on ongoing maintenance of the Chinese website (www.fijiislands.com.cn); hosting media outlets under the international media and television programme; invitation of Chinese media and trade attending the annual Bula Fiji Tourism Exchange; attendance at key promotional shows in China; publication of Chinese brochures and pamphlets; media and joint cooperatives with outbound operators to target the prime holiday season in China. The closure of the Tourism Fiji Shanghai Office in 2012 subsequently resulted in the shutdown of the Tourism Fiji Chinese website which would have affected the consistency in marketing.

For Fiji, in particular, the traditional markets of Australia and New Zealand provide more repeat visitors relative to China, where most of the Chinese tourists to Fiji are first-time visitors. Therefore, in terms of destination marketing, Fiji needs to facilitate and spearhead more focused marketing on specific target markets in the Chinese outbound market. A study by Sofield and Li (2011) indicated that many of the marketing initiatives that work for Fiji's current source markets will fail to hit the mark in China and that it is critical to have an understanding of the key aspects of the Chinese value system, the culture, philosophy, and world view of the Chinese people. One such example is that the Chinese enjoy nature, but on their terms, meaning that nature needs to be packaged, tamed, and beautified with landscaped gardens, concrete paths, and stairs for trails that run through forests. Whilst Westerners may view this as spoiling nature, the Chinese however view this as men being in harmony with nature.

There is also evidence from tourism destinations such as the Maldives which has experienced problems in dealing with the Chinese outbound market. A result of the limited planning and preparation for Chinese tourists to the Maldives led to many resort operators being ignorant of how to cater to the different needs and expectations of Chinese visitors and the lack of Chinese brochures, signage, speakers, and cuisine led to misunderstandings and irritation for both the host and tourists (Sofield & Li, 2011)

Destination branding is critical to distinguish destinations and to convey a positive message that will motivate tourists to visit the destination (Middleton, 1994). Fiji's previous branding slogan 'Fiji me' and current branding slogan 'Fiji', where happiness finds you are tailored to all target markets, whether it is for the wedding and honeymoon market, adventure market, or family market. In addition, this same branding is used for all Fiji's source markets, whether it is Australia, New Zealand, Korea, Russia, or China. Therefore, different branding slogans must be re-created specifically for the Chinese market and include a positive message that is tailored to different sub-groups.

Conclusion

This chapter suggests that if tourism stakeholders in Fiji are to target the Chinese outbound market, then their staff, especially front-line staff must be aware and have an overall understanding of the different aspects of the Chinese culture. The hosts' ability to respond effectively to a culturally different tourist is an important element determining positive tourist holiday experiences and satisfaction. What was also highlighted in the interviews with key tourism stakeholders is that if tourism operators are keen to target this market, they need to ensure that they have the budget and the commitment to be consistent in targeting this growing emerging market. More importantly, marketing strategies need to be aligned to the appropriate market segment and via relevant media.

The findings from this study indicate a need to regulate the Chinese inbound market to Fiji. There have been past experiences with tourism destinations such as Maldives, Seychelles, Mauritius, Australia, and New Zealand that have opened their doors to the Chinese outbound market and are now experiencing challenges in dealing with a mass market that is difficult to control and is affecting repeat visitors from traditional markets. As China is an emerging market for Fiji, tourism stakeholders must take heed of lessons from matured destinations and try and regulate this market. In addition, tourism stakeholders can also consider targeting the Chinese outbound market only during the traditional low seasons; or to assign a specific number of rooms for the Chinese market so as not to impact existing guests from traditional markets. Furthermore, destination marketing and promotion play a key role in how Fiji positions itself in the Chinese outbound market, and as the Ministry of Tourism and Tourism, Fiji's main prerogative is to increase tourist numbers; the onus then rests on the properties (i.e. hotels and resorts) to position Fiji to target luxury Chinese travellers and also to implement strategies in terms of controlling and regulating the Chinese inbound market to Fiji.

In addition, whilst many tourism developers and operators are following good practices, Fiji lacks the framework to ensure that such practices are adopted across the tourism industry. Much of the policy, legislation, and regulation needed to ensure that good practice is not implemented or enforced. Tourism practitioners in Fiji need to acknowledge and recognize that the possibility of mass numbers of Chinese tourists can adversely impact tourists from traditional markets and also have social impacts on the destination if proper preparation and adequate planning are not taken into account by Government and key tourism industry stakeholders.

Whilst adequate knowledge and viewpoints were extracted from the six tourism stakeholders, it would have been valuable to have more stakeholder representation that contributes to the tourist experience such as retail shopping outlets, restaurants, excursions, and tour operators to obtain additional insight into the Chinese tourist to Fiji.

Acknowledgement: This chapter is based on the first author's Master's thesis entitled "Diversification of Fiji's tourist market: A study of Chinese tourists to Fiji" (2015), University of the South Pacific, Fiji.

References

Agrusa, J., Kim, S. S., & Wang, K.-C. (2011). Mainland Chinese tourists to Hawaii: Their characteristics and preferences. *Asia Pacific Journal of Tourism Research, 28*(3), 261–278. doi:10.1080/10548408.2011.562853

Baseball, E. (2010). China TV outlet to visit for the campaign. http://www.fijitimes.com/story.aspx?id=153335

Bernard, H. (2001). *Research methods in anthropology: Qualitative and quantitative approaches.* Lanham, MD: AltaMira Press.

Boycoff, P. (2014, January 20). U.S to be a top destination for Chinese tourists. *CNN Business.* Retrieved August 21, 2013, from http://money.cnn.com/2014/01/20/news/economy/china-tourism-us/

China International Travel Monitor. (2014). China International Travel Monitor 2014. http://press.hotels.com/content/themes/CITM/CITM_UK_PDF_2014.pdf

Deery, M., Jago, L., & Fredline, L. (2012). Rethinking social impacts of tourism research: A new research agenda. *Tourism Management, 33*(1), 64–73. doi:10.1016/j.tourman.2011.01.026

Fiji Bureau of Statistics. (2013). Provisional visitor arrivals. https://www.statsfiji.gov.fj/latest-releases/tourism-and-migration/visitor-arrivals

Fiji Times. (2009). Tourism Fiji officially salutes Asian market growth. http://www.fijitimes.com/story.aspx?id=127691

Foley, A., & Fahy, J. (2004). Incongruity between expression and experience: The role of imagery in supporting the positioning of a tourism destination brand. *Journal of Brand Management, 11*(3), 209–217. doi:10.1057/palgrave.bm.2540167

Fraser, B. (1990). Perspectives on politeness. *Journal of Pragmatics, 14*(2), 219–236. doi:10.1016/0378-2166(90)90081-n

Friedman, A. L., & Miles, S. (2006). *Stakeholders: Theory and practice.* Oxford: Oxford University Press.

Fyall, A., & Leask, A. (2006). Destination marketing: Future issues—Strategic challenges. *Tourism and Hospitality Research*, *7*(1), 50–63. doi:10.1057/palgrave.thr.6050029

García, J. A., Gómez, M., & Molina, A. (2012). A destination-branding model: An empirical analysis based on stakeholders. *Tourism Management*, *33*(3), 646–661. doi:10.1016/j.tourman.2011.07.006

Gong, J., Detchkhajornjaroensri, P., & Knight, D. W. (2018). Responsible tourism in Bangkok, Thailand: Resident perceptions of Chinese tourist behaviour. *International Journal of Tourism Research*, *21*, 1–13. doi:10.1002/jtr.2256

Hall, M., & Page, S. (1996). *Tourism in the South Pacific: Issues and cases*. London: International Thomson Business Press.

Hanqin, Z. Q., & Lam, T. (1999). An analysis of Mainland Chinese visitors' motivations to visit Hong Kong. *Tourism Management*, *20*(5), 587–594. doi:10.1016/s0261–5177(99)00028-x

Harrison, D. (2003). Themes in Pacific Island tourism. In D. Harrison (Ed.), *Pacific Island Tourism* (pp. 1–24). New York: Cognizant Communication Corporation.

Harrison, D., & Pratt, S. (2010). Political change and tourism: Coups in Fiji. In R. Butler & W. Suntikul (Eds.), *Tourism and political change* (pp. 160–174). Oxford: Goodfellow Publishers.

Heath, E., & Wall, G. (1992). *Marketing tourism destinations: A strategic planning approach*. New York: Wiley.

Ho, E. (2011, September 26). Australia tells Chinese tourists: Welcome! (and bring your checkbooks). *Time*. Retrieved August 19, 2013, from http://content.time.com/time/world/article/0,8599,2094863,00.html

Jang, S. (Shawn), Yu, L., & Pearson, T. (2003). Chinese travellers to the United States: A comparison of business travel and visiting friends and relatives. *Tourism Geographies*, *5*(1), 87–108. doi:10.1080/1461668032000034097

Kanemasu, Y. (2013). Social construction of touristic imagery: Case of Fiji. *Annals of Tourism Research*, *43*, 456–481. doi:10.1016/j.annals.2013.06.004

Kim, S., Guo, Y., & Agrusa, J. (2012). Preference and position analyses f overseas destinations by mainland Chinese outbound pleasure tourists. *Journal of Travel Research*, *44*, 212–220.

Lal, R. (2011, November 11). Need to entice Chinese tourists. *Fiji Sun*, p. 7.

Li, X. (Robert), Lai, C., Harrill, R., Kline, S., & Wang, L. (2011). When east meets west: An exploratory study on Chinese outbound tourists' travel expectations. *Tourism Management*, *32*(4), 741–749. doi:10.1016/j.tourman.2010.06.009

Lo, A., & Lam, T. (2004). Long-haul and short-haul outbound all-inclusive package tours. *Asia Pacific Journal of Tourism Research*, *9*(2), 161–176. doi:10.1080/1094166042000233694

Middleton, V. (1994). *Marketing in travel and tourism*. Oxford: Reed Elsevier.

Ministry for Industry, Trade & Tourism. (2020). *Fijian Tourism 2021*. https://mitt.gov.fj/wp-content/uploads/2019/04/FT2021.pdf

Ministry of Tourism. (2007). *Fiji's tourism development plan 2007–2016*. Suva: Department of Ministry of Tourism & Environment.

Morrison, A. M. (2013). *Marketing and managing tourism destinations*. New York: Routledge.

Narayan, P. K. (2000). Fiji's tourism industry: A SWOT analysis. *Journal of Tourism Studies*, *11*(2), 15–24.

Native Land Trust Board. (2006). *A tourism policy for native land 2006–2010*. Suva: National Land Trust Board.

Opperman, M., & Chon, K. (1997). *Tourism in developing countries*. London: International Thomson Business Press.

Pan, G. W., & Laws, E. (2007). Tourism development of Australia as a sustained preferred destination for Chinese tourists. *Asia Pacific Journal of Tourism Research*, *8*(1), 37–47. doi:10.1080/10941660308725454

Patton, M. Q. (2002). *Qualitative research and evaluation methods*. Thousand Oaks, CA: Sage.

Rao, M. (2002). Challenges and issues for tourism in the South Pacific island states: The case of the Fiji Islands. *Tourism Economics*, *8*(4), 401–429. doi:10.5367/000000002101298205

Reddy, S. (2010, August 14). National update and a path to sustainable tourism growth. *[Conference presentation]. Fiji Tourism Forum 2009, Suva Fiji*. http://www.bis.org/review/r090819d.pdf

Reisinger, Y., & Turner, L. W. (2002). Cultural differences between Asian tourist markets and Australian hosts, Part 1. *Journal of Travel Research*, *40*(3), 295–315. doi:10.1177/004728750204000308

Samy, J. (1980). Crumbs from the table: The workers share in tourism. In F. Rajotte. (Ed.), *Pacific tourism: As islanders see it* (pp. 67–82). Suva: University of the South Pacific.

Sautter, E. T., & Leisen, B. (1999). Managing stakeholders a tourism planning model. *Annals of Tourism Research*, *26*(2), 312–328. doi:10.1016/s0160–7383(98)00097-8

Scott, N., Cooper, C., & Baggio, R. (2008). Destination networks: Four Australian cases. *Annals of Tourism Research*, *35*(1), 169–188. doi:10.1016/j.annals.2007.07.004

Serelini, K. (2010, February 20). Chinese tourists are good for Fiji. *Fiji Sun*, p. 26.

Sofield, T. (2003). *Empowerment for sustainable tourism development*. London: Elsevier Science.

Sofield, T., & Li, S. (2011). Tourism governance and sustainable national development in China: A macro-level synthesis. *Journal of Sustainable Tourism*, *19*(4–5), 501–534. doi:10.1080/09669582.2011.571693

Song, H., & Lin, S. (2009). Impacts of the financial and economic crisis on tourism in Asia. *Journal of Travel Research*, *49*(1), 16–30. doi:10.1177/0047287509353190

South Pacific Tourism Organisation. (2009). *Annual review of visitor arrivals in the Pacific*. https://corporate.southpacificislands.travel/wp-content/uploads/2017/02/2016-Annual-Visitor-Arrivals-ReviewF.pdf

Sparks, B., & Pan, G. W. (2009). Chinese Outbound tourists: Understanding their attitudes, constraints, and use of information sources. *Tourism Management*, *30*(4), 483–494. doi:10.1016/j.tourman.2008.10.014

The Jakarta Post. (2011). Bali's tourism industry must diversify its Industry groups. Retrieved March 12, 2012, from http://thejakartapost.com/news/2011/03/28/bali

Tse, T. S. M., & Hobson, J. S. P. (2008). The forces shaping China's outbound. *Journal of China Tourism Research*, *4*(2), 136–155. doi:10.1080/19388160802279459

United Nations World Tourism Organization. (2011). *International tourists to hit 1.8 billion by 2030*. http://media.unwto.org/ar/node/33508

Vernon, J., Essex, S., Pinder, D., & Curry, K. (2005). Collaborative policymaking: Local sustainable projects. *Annals of Tourism Research*, *32*(2), 325–345. doi:10.1016/j.annals.2004.06.005

Waligo, V. M., Clarke, J., & Hawkins, R. (2013). Implementing sustainable tourism: A multi-stakeholder involvement management framework. *Tourism Management*, *36*, 342–353. doi:10.1016/j.tourman.2012.10.008

Wang, Y., & Krakover, S. (2008). Destination marketing: Competition, cooperation or competition?. *International Journal of Contemporary Hospitality Management*, *20*(2), 126–141. doi:10.1108/09596110810852122

Weiermair, K. (2000). Tourists' perceptions towards and satisfaction with service quality in the cross-cultural service encounter: Implications for hospitality and tourism management. *Managing Service Quality: An International Journal*, *10*(6), 397–409. doi:10.1108/09604520010351220

World Wide Fund. (2003). *WWF Annual Review 2003*. https://wwf.panda.org/?12663/WWF-Annual-Review-2003

Xie, Y., & Li, M. (2009). Development of China's outbound tourism and the characteristics of its tourist flow. *Journal of China Tourism Research*, *5*(3), 226–242. doi:10.1080/19388160903149965

Xinhua. (2007, August 19). Macao expands the Asian tourism market to boost diversification. *People's Daily China*. Retrieved April 15, 2012, from http://english.people.com.cn/90001/90778/6242975.html

Zhang, H., & Heung, V. (2002). The emergence of the mainland Chinese outbound. *International Journal for Innovation Education and Research*, *1*(4), 23–29.

Zhang, X. (2013). A study of Chinese tourists' bad behaviors. *International Journal for Innovation Education and Research*, *1*(4), 23–29.

Zhao, C. (2006). *New Zealand's destination image and the Chinese outbound market: A comparative study between Beijing (North) and Guangdong (South) markets* [Unpublished doctoral dissertation]. Auckland University of Technology.

Zhou, L., King, B., & Turner, L. (1998). The China outbound market: An evaluation of key constraints and opportunities. *Journal of Vacation Marketing*, *4*(2), 109–119. doi:10.1177/135676679800400203

9 Identifying the key factors influencing service quality among Chinese GTPs in New Zealand

Shaohua Yang, Salmi Mohd Isa and Yedan Fan

Introduction

The Approved Destination Status scheme (ADS) was introduced in 1999 to support the growth of Chinese tourism market to New Zealand (Tourism New Zealand, 2016). This policy allows Chinese tourists to visit New Zealand as part of a tour group on an ADS visa. Recent years have seen a rapid growth in Chinese inbound market for New Zealand. Consequently, there emerges measures to attract Chinese tourists, and China becomes New Zealand's largest tourism market in the world (Tourism New Zealand, 2018).

In February 2019, more than 50,900 Chinese tourists have come to New Zealand (Yang et al., 2021). Official statistics estimated Chinese tourists' annual expense of NZ$4.3 billion ($2.92 billion) for travelling to New Zealand by 2023 (GMA, 2017). However, growing popularity of free and independent travel among Chinese to New Zealand, the group tour still accounts for about 46% of Chinese travellers by March 2016 (Tourism New Zealand, 2016). Overall, Chinese is the largest source of visitors to New Zealand, and its volume and expenditure continue to grow, which thus attracts relevant parties' attention.

New Zealand has several elements that help facilitate Chinese tourists (Yang et al., 2020). For instance, Ryan and Mo (2002) summarised Chinese tourists' multiple motivations to travel to New Zealand. First, they arrived in New Zealand to conduct "social investigations" such as visiting friends/relatives and researching business market along with local educational system. Second, Chinese' special interest in New Zealand originates from their perception of New Zealand as a safe, clean, unpolluted and pleasant tourism destination. Third, various vocational destinations there also offer them an escape from daily life and to pursue an adventurous, new or relaxing experience. The last motivation concerns their curiosity about exploring the unknown territories. Hence, based on above reasons, New Zealand tourism industry has a competitive advantage in the Chinese market.

China's three popular travel modes are independent tour, group tour and group package tour (GPTs). For the Chinese outbound tourists, GPTs are the primary modes of outbound travel and are likely to retain popularity for

DOI: 10.4324/9781003121329-11

international travel (Wang & Sheldon, 1996). Middleton and Clarke (2012) defined GPTs as standardised and repeatable bundle offers sold at single prices, which often involve transportation, accommodation, food, attractions and services. It has been revealed that GPTs are intriguing due to its financial profits (i.e. value for money and reasonable price), convenience and less uncertainties and risk perceived by users (Kim et al., 2009; Wong & McKerracher, 2012). Many Chinese tourists select GPTs as their preferred model for tourism. Travelling on GPTs enables Chinese to overcome language barriers, a comparatively immature international market to flourish, and it also matches China's collectiveness (Li et al., 2011). The GPTs programme offers a good platform for Chinese to develop their networking with other Chinese.

Although New Zealand has implemented successful marketing strategies that have attracted a large number of Chinese group tourists, several shortcomings remain that affect Chinese group tourists' destination image of New Zealand adversely. In New Zealand, shopping is a popular activity among Chinese tourists, especially for souvenirs to bring home to friends or others (Becken, 2003; Han, 2018; Pan & Laws, 2002). Chinese tourists prefer purchasing health products because they believe that New Zealand's environments are very clean and wholesome. Zhao (2006) found that South Chinese are concerned most about health, stress, escaping life stressors and as such seek to purchase health products while they are travelling to New Zealand. Chinese travellers spend extremely large amounts of money on shopping (Petersen, 2009; Yu, 2012). This shopping behaviour has also been acknowledged universally.

In New Zealand, the NBR contributor (Bond, 2013) stated that Chinese love shopping and that it is a part of their DNA. This shopping behaviour complies with the gift-giving culture in China. It is common for Chinese to buy gifts for their friends, family, colleagues and commercial partners who are essential for them to build personal network. Hence, shopping is an essential activity for Chinese tourists, particularly souvenirs to give as gifts to family, friends and others (Pan & Laws, 2002). Giving such gifts is also a status symbol in Chinese culture (Guo et al., 2007).

However, due to an in-depth understanding of Chinese shopping behaviour, travel agents and tour leaders seek to maximise their profits from tourist spending. To achieve this, tour agents/tours have been found to force Chinese tourists in New Zealand to go shopping (Devereux, 2017). When Chinese tourists select GPTs to travel to New Zealand, most packing itineraries designed by tour operator emphasise more shopping destinations. As such, Chinese tourists' overall service quality and satisfaction can be affected by tour shopping arrangements (Chen et al., 2018).

Besides that, other factors can also influence the service quality among Chinese GPTs in New Zealand. For instance, the significant difference in culture, food standard and accommodation standard between Western countries and China could influence Chinese tourists' expectations. For the service providers, they can find it challenging to accommodate the Chinese tourists' preference for their food and luxury hotels especially for luxurious GPTs (Jørgensen et al., 2018; Ng et al., 2007). An (2018) found that service providers (e.g., hotel and food) find it difficult to meet and exceed Chinese tourists' expectations which can adversely affect overall service quality and travel experience.

Additionally, New Zealand's Ministry of Tourism want to foster confidence in the travel services and products offered by operators for Chinese GPTs to New Zealand (New Zealand Tourism, n.d.). Nonetheless, there is information mismatch between tour packages and Chinese tourists. Travel agencies offer cheap tours in their attempt to reduce costs as much as possible and want to recoup their costs or maximise profits from commissions in partner shops. The adverse consequences of such behaviour are clear of tourists' experience.

Based on the above discussion, Chinese GPTs experience low levels of service quality in New Zealand. As such, GPTs often hurt Chinese tourists. To remedy this situation, tourism marketers in New Zealand should focus not only on effective marketing strategies to attract Chinese tourists from China but should also focus on offering effective service strategies to maintain a high-quality tourist experience and high levels of tourist satisfaction. Satisfied tourists are likely to return and make positive word of mouth recommendations about tourism companies (Duke & Persia, 1996).

In the wider business and tourism landscape, quality service has become an increasing demand, which is known to businessmen (Schwartz, 2007; Ting, 2004; Velázquez et al., 2011) particularly those engaging in tourism industry (Ayeh & Chen, 2011; Mak et al., 2011). To achieve sustainable tourism in the long run, Chinese tourism operators should become quality service-oriented (Chen et al., 2018). An important strategy for achieving this is to diagnose service shortfalls. Therefore, in this chapter, we identify several service shortfalls that influence service quality among Chinese GPTs in New Zealand. Subsequently, we propose several recommendations for improvements. It is expected that these recommendations will be beneficial for tourism stakeholders.

Literature review

Scholars suggest using SERVQUAL as a marketing theory to measure the quality of services (Boulding et al., 1993; Parasuraman et al., 1985, 1988, 1991) and identify acceptable service standards across the various industries (Cronin & Taylor, 1994; Parasuraman et al., 1988). However, SERVQUAL fails to paint an overall picture of Chinese GPTs service (Chen et al., 2018). It is unlikely that the categorisation in general service can be directly applied to the complicated service as GPTs (Wang et al., 2000, 2007). Therefore, it is necessary to explore dimensions of service models other than SERVQUAL. A study indicated that the tour package has an indirect impact on service quality among Chinese GPTs (Chen et al., 2015). The tour package cannot satisfy Chinese tourists' expectations and can hardly provide quality services. In consumer behaviour, the selection of product bundles like a package tour can be traced to the theory of characteristics defined by Lancasters (1966a, 1966b). According to this theory, utility is based on consumers' perception of something as good rather than the good itself (Lancaster, 1966a, 1966b).

Group tour packages and service quality

In the tourism context, characteristics theory elaborates on tourists' responses to different features of a tour package. For instance, a model on tourists' choices underscores two complementary features of a destination which include "historical attractions" versus "scenic beauty" or "facilities" versus "attractions" (Papatheodorou, 2001; Rugg, 1973). However, there is still a lack of focus on how tour operators decide to bundle a range of travel products and services in a tour package, which may, in turn, influence service quality and consumption (Chen et al., 2015).

From the alternative tourism perspective, Chinese group tourists travel to destinations for different purposes. The same package tour group consists of tourists with different levels of income and social status resulting in different expectations of service quality. As such, package tours often fail to provide good services to Chinese tourists. They might dent the demand as it would be hard for consumers to evaluate the exact value of every single component in those products (Gourville & Soman, 2005; Soman & Gourville, 2001). When the package tour cannot meet Chinese GPTs expectation, Chinese tourists can feel frustrated and cheated. This issue is of critical importance for the package tour programme because tour operators designed package tours that are incompatible with Chinese tourists' values and expectations.

Information asymmetry and service quality

When there is a mismatch between tour package and Chinese GPTs, it results in a negative impact on service quality. Aggressive tourism promotion in the Chinese tourism market has resulted in information overload and asymmetry. Tourists select information sources online to pre-evaluate the package tour. How clear information tourists receive from tour operators can influence their purchasing choices of package tours (Gursoy & Chen, 2000; Gursoy & McCleary, 2004; Xiang & Gretzel, 2010). When promoting package tours, tour operators use product handling strategies which help to maximise profits while charging low prices to Chinese GPTs.

However, tour operators never mention the service standard which could reduce due to low-priced tour packages. Chinese GPTs usually purchase this type of package. Wealthy Chinese tourists can feel particularly cheated and dissatisfied by such package tours. Studies have warned about issues such as adverse selection and moral hazard in the tourism market, for which tourists experience low-quality service and dissatisfaction (Candela & Figini, 2012; Caserta & Russo, 2002; Keane, 1996, 1997; Reece, 2009). As such, since information asymmetry between package tour and tourists affects service quality, service quality is determined by the degree of mismatch between Chinese GPTs and package tour.

The role of the tour leader and service quality

For the Chinese GPTs, a good tour leader is indispensable for quality service (Wang et al., 2000). His/her performance directly influences the travelling experience and tourists' satisfaction (Yu et al., 2002). During the package tour, the tour leaders supply multiple services such as ensuring tourists' security and safety, offering them valuable information, facilitating group interaction, working as surrogate parent, pathfinder, mentor, leader, mediator and entertainer (Wang & Davidson, 2010; Wang et al., 2015). In package tour studies, Wang et al. (2000) explored the critical incidents related to the tour leaders that contribute to dis/satisfaction. These incidents occurred most frequently in the areas of shopping, extra tours itineraries (self-sponsors) and airport/aeroplane services. Due to intense competition, tour leaders rely on commission from shopping and extra tour itineraries to increase their income.

Most tour leaders in New Zealand are self-employed and force Chinese GPTs to go shopping. In other words, they are not employed by any segment of the tourist industry. Their income is based on commission leading them to collaborate with local souvenir stores, hotels, restaurants and other entertainment services. Ryan et al. (2013) launched a summary report and indicated that Chinese tourists expressed distrust in guides concerning shopping tours. The report revealed that tour guides hold diverse attitudes to shopping tours, which range from dislike to regarding them as a way to earn money. Additionally, in New Zealand tour, leaders sometime manipulate the shopping arrangement and deliver misleading information to Chinese GPTs. For instance, Chinese tourists are encouraged to undertake their shopping earlier in the tour because the tour leaders told Chinese that it was not safe if we walk outside the hotel in the evening. When doing this, they prevent Chinese tourists from opportunities to visit other shops.

Furthermore, King et al. (2006) concluded with three main points which can hurt and damage service quality among Chinese GPTs in New Zealand: first, shopping arrangements that are not competitive include extra or covert commissions, under-cost packages, exclusive dealings and exercise of influence; second, false or misleading information such as misrepresentation of travel components, quality of accommodation and freely available services; and third, low-quality service caused by poorly designed packages and employees with insufficient skills.

The large body of literature demonstrates that the role of the tour leader influences service quality for tourists. For instance, factors such as commission tours have increased tourists' dissatisfaction and complaints about unfair pricing of goods at scenic spots, changing itineraries for the destination, visiting shopping venues to tour operators' advantage, misleading tourists to purchase expensive goods, misrepresentation of information about destinations by tour leaders and compulsive shopping (Dwyer et al., 2005, 2007; Keating, 2009; Prideaux et al., 2006; Wang et al., 2015; Zhang & Murphy, 2009). Given their role, service quality is determined by the role of tour leaders.

Conclusion

Based on the discussion, it can be concluded that alternative key elements such as package tour, information asymmetry and the role of tour leaders are antecedents of service quality among Chinese GPTs. On top of understanding, the dimensions of the SERVQUAL the model, tourism stakeholders should also explore package tours, information asymmetry and the role of tour leaders which exert salient influence on Chinese GPTs in New Zealand. New Zealand tourism stakeholders might need to monitor service performance of these elements to help to develop a sustainable package tourism industry.

For the package tour, tour operators should conduct more marketing research to analyse Chinese tourists' needs, expectations and tour values. Categorising the demographic profile of Chinese tourists can also enhance service quality. For example, when the Chinese GPTs have a preference for shopping, then the package tour can be designed based on a shopping-oriented trip. For those Chinese tourists interested in seeing attractions and experiencing local customs, shopping should be minimised unless otherwise expressed by the tourists. This would also help the tourism market to select effective marketing segmentation strategies. Tour operators can explore other profitable business models. This might be achieved through appropriate price levels and adjusting package tours.

In service quality, it is illustrated by the expectancy disconfirmation theory that the mismatch between expectation and actual experience entails dissatisfaction (Wang & Davidson, 2010). In relation to the information asymmetry between package tour and GPTs, travel agents in China should focus on ethical promotional approaches. All the information related to package tours and background of tour leaders in New Zealand should be communicated clearly to potential Chinese tourists. Existing marketing and promotional efforts should be more realistic in their imagery and expectations. Otherwise, the package tour will not meet Chinese GPTs expectations resulting in a problem of service quality. Based on above these notions, tourism marketers in New Zealand could also build a good match of what is offered in package tours and what travellers expect to receive regarding service quality.

As for tour leaders, one approach to address the service gaps between tour leaders and Chinese GPTs is heavier relegation by New Zealand tourism authorities. It would be beneficial to Chinese GPTs if the government can play a stronger role in eliminating unethical commission-based tourisms. Optional activities during tours such as shopping should be in compliance with the trip contract in strict sense. New Zealand and Chinese tourism authorities could work together to take action to enforce the law and conduct evaluations of its effectiveness. Regulations and licences relevant to inbound and outbound tour business and its governance should be introduced and enforced more strictly. For instance, all tour leaders are required to comply or otherwise risk losing their licence for breaching the law. It is assumed that the laws of competition can reward and motivate tour leaders to offer high-quality service by the method of positive word of mouth communication and repurchasing package tour. Setting the point system for tour licensing, points can be deducted if tour leaders provide poor service quality to Chinese GPTs.

Moreover, at the national level, tourism policymakers in New Zealand can adjust its human resource strategies. The pay structure model including pay by leaders, pay by followers and pay by performance should be adopted for tour leaders and eliminate the policy that tour leaders should be self-employed. As for relevant organisations in New Zealand, they can consider employing personnel to recruit, train and monitor tour leaders' performance. The tourism industries might need to recruit and train tour leaders with competent interpersonal and problem-solving skills.

The limitations of this chapter should be addressed for further work. This chapter only covered several factors influencing service quality among Chinese GPTs. Further studies could explore better accreditation of tour leaders and the development of trade arrangements between China and New Zealand. As the chapter based only on the discussion of previous studies, there is no empirical evidence to support these key concepts (tour package, information asymmetry and tour leader). It is suggested that further studies focus on scale development and test these dimensions in a conceptual framework by providing empirical evidence.

References

An, T. B. (2018). *Guest satisfaction in New Zealand luxury hotels: A study of reviews from TripAdvisor* [Master dissertation, Auckland University of Technology]. Auckland University of Technology Library. https://openrepository.aut.ac.nz/handle/10292/11709

Ayeh, J. K., & Chen, R. X. (2011). 'How's the service?' A study of service quality perceptions across sectors and source markets. *International Journal of Tourism Research*, *15*(3), 241–260. doi:10.1002/jtr.1864

Becken, S. (2003). Chinese tourism to New Zealand: Discussion paper. *Landcare Research*, 1–15. https://www.landcareresearch.co.nz/uploads/public/researchpubs/becken_Chinese_tourism.pdf

Bond, G. (2013, September 7). China moves to end jacked-up NZ bus tour shopping stops. *National Business Review*. Retrieved January 15, 2019, from https://www.nbr.co.nz/article/chinese-travelers-shunning-tour-bus-b-p-145297

Boulding, W., Kalra, A., Staelin, R., & Zeithaml, V. A. (1993). A dynamic process model of service quality: From expectations to behavioral intentions. *Journal of Marketing Research*, *30*(1), 7–27. doi:10.2307/3172510

Candela, G., & Figini, P. (2012). *The economics of tourism destinations*. Berlin: Springer.

Caserta, S., & Russo, A. P. (2002). More means worse: Asymmetric information, spatial displacement and sustainable heritage tourism. *Journal of Cultural Economics*, *26*(4), 245–260.

Chen, H., Weiler, B., & Young, M. (2018). Examining service shortfalls. A case study of Chinese group package tours to Australia. *Journal of Vacation Marketing*, 1–18. doi:10.1177/1356766717750420

Chen, H., Weiler, B., Young, M., & Lee, Y. L. (2016). Conceptualizing and measuring service quality: Towards consistency and clarity in its application to travel agencies in China. *Journal of Quality Assurance in Hospitality & Tourism*, *17*(4), 516–541. doi:10.1080/1528008x.2015.1133365

Chen, Y., Schuckert, M., Song, H., & Chon, K. (2015). Why can package tours hurt tourists? Evidence from China's tourism demand in Hong Kong. *Journal of Travel Research*, *55*(4), 427–439. doi:10.1177/0047287515612597

Cronin, J. J., & Taylor, S. A. (1994). SERVPERF versus SERVQUAL: Reconciling performance-based and perceptions-minus-expectations measurement of service quality. *Journal of Marketing*, *58*(1), 125–131. doi:10.1177/002224299405800110

Devereux, C. (2017, April 4). Data dump: How Chinese laws affect New Zealand tourism. *The Register*. Retrieved January 28, 2019, from https://theregister.co.nz/news/2017/04/data-dump-chinese-new-zealand-tourism

Duke, C. R., & Persia, M. A. (1996). Consumer-defined dimensions for the escorted tour industry segment: Expectations, satisfactions, and importance. *Journal of Travel & Tourism Marketing*, *5*(1–2), 77–99. doi:10.1300/j073v05n01_06

Dwyer, L., King, B., & Prideaux, B. (2005). Effect of unethical business practices on Australian inbound package tourism: Study into inbound practices affecting product quality in Australian inbound tourism. In *Proceedings of the Council for Australian University Tourism and Hospitality Education Conference, from CAUTHE 2005: Sharing Tourism Knowledge, Alice Springs, NT, Australia*. https://researchonline.jcu.edu.au/7786/

Dwyer, L., King, B., & Prideaux, B. (2007). The effects of restrictive business practices on Australian inbound package tourism. *Asia Pacific Journal of Tourism Research*, *12*(1), 47–64. doi:10.1080/10941660601036009

GMA, O. (2017, September 22). Chinese Tourists in New Zealand. *Chinese Tourists Agency*. Retrieved January 17, 2019, from http://chinesetouristagency.com/chinese-tourists-in-new-zealand/

Gourville, J. T., & Soman, D. (2005). Over choice and assortment type: When and why variety backfires. *Marketing Science*, *24*(3), 382–395.

Guo, Y., Seongseop Kim, S., & Timothy, D. J. (2007). Development characteristics and implications of mainland Chinese outbound tourism. *Asia Pacific Journal of Tourism Research*, *12*(4), 313–332. doi:10.1080/10941660701760995

Gursoy, D., & Chen, J. S. (2000). Competitive analysis of cross cultural information search behavior. *Tourism Management*, *21*(6), 583–590. doi:10.1016/s0261-5177(00)00005-4

Gursoy, D., & McCleary, K. (2004). An integrative model of tourists' information search behaviour. *Annals of Tourism Research*, *31*(2), 353–373. doi:10.1016/j.annals.2003.12.004

Han, P. (2018). *Culture and souvenir shopping. How do Chinese cultural values play a role in outbound Chinese tourists' souvenir shopping?* [Master's thesis, UiT The Arctic University of Norway]. UiT The Arctic University of Norway. https://hdl.handle.net/10037/14846

Jørgensen, M. T., Law, R., & King, B. E. (2018). Beyond the stereotypes: Opportunities in China inbound tourism for second-tier European destinations. *International Journal of Tourism Research*, *20*(4), 488–497. doi:10.1002/jtr.2198

Keane, M. J. (1996). Sustaining quality in tourism destinations: An economic model with an application. *Applied Economics*, *28*(12), 1545–1553. doi:10.1080/000368496327525

Keane, M. J. (1997). Quality and pricing in tourism destinations. *Annals of Tourism Research*, *24*(1), 117–130. doi:10.1016/s0160-7383(96)00034-5

Keating, B. (2009). Managing ethics in the tourism supply chain: The case of Chinese travel to Australia. *International Journal of Tourism Research*, *11*(4), 403–408. doi:10.1002/jtr.706

Kim, J., Bojanic, D. C., & Warnick, R. B. (2009). Price bundling and travel product pricing practices used by online channels of distribution. *Journal of Travel Research*, *47*(4), 403–412. doi:10.1177/0047287508328658

King, B., Dwyer, L., & Prideaux, B. (2006). An evaluation of unethical business practices in Australia's China inbound tourism market. *International Journal of Tourism Research*, *8*(2), 127–142. doi:10.1002/jtr.563

Lancaster, K. J. (1966a). A new approach to consumer theory. *Journal of Political Economy*, *74*(2), 132–157.

Lancaster, K. J. (1966b). Change and innovation in the technology of consumption. *American Economic Review*, *56* (1/2), 14–23. doi:10.2307/1821263

Li, M., Zhang, H., Mao, I., & Deng, C. (2011). Segmenting Chinese outbound tourists by perceived constraints. *Journal of Travel & Tourism Marketing*, *28*(6), 629–643. doi:10.1080/10548408.2011.598747

Mak, A. H. N., Wong, K. K. F., & Chang, R. C. Y. (2011). Critical issues affecting the service quality and professionalism of the tour guides in Hong Kong and Macau. *Tourism Management*, *32*(6), 1442–1452. doi:10.1016/j.tourman.2011.01.003

Middleton, V. T. C., & Clarke, J. R. (2012). *Marketing in travel and tourism*. Oxford: Routledge.

New Zealand Tourism. (n.d.). China group tour. *100% Pure New Zealand*. Retrieved February 22, 2019, from https://traveltrade.newzealand.com/en/working-with-us/china-group-tours/

Ng, S. I., Lee, J. A., & Soutar, G. N. (2007). Tourists' intention to visit a country: The impact of cultural distance. *Tourism Management*, *28*(6), 1497–1506. doi:10.1016/j.tourman.2006.11.005

Pan, G. W., & Laws, E. (2002). Tourism marketing opportunities for Australia in China. *Journal of Vacation Marketing*, *8*(1), 39–48. doi:10.1177/135676670200800105

Papatheodorou, A. (2001). Why people travel to different places. *Annals of Tourism Research*, *28*(1), 164–179. doi:10.1016/s0160-7383(00)00014-1

Parasuraman, A., Berry, L. L., & Zeithaml, V. A. (1991). Understanding customer expectations of service. *Sloan Management Review*, *32*(2), 39–48.

Parasuraman, A., Zeithaml, V. A., & Berry, L. L. (1985). A conceptual model of service quality and its implications for future research. *Journal of Marketing*, *49*(4), 41–50. doi:10.1177/002224298504900403

Parasuraman, A., Zeithaml, V. A., & Berry, L. L. (1988). SERVQUAL: A multiple-item scale for measuring consumer perceptions of service quality. *Journal of Retailing*, *64*(1), 12–37.

Petersen, S. L. (2009). *Chinese tourists' travel motivation and their resultant preferred travel activities.* Retrieved May 28, 2012, from http://www.odense.dk/Topmenu/Borger/ByMiljoe/Odense%20i%20Verden/EXPO %202010/Presse/~/media/BKF/Bymiljø/Expo%202010/Nyheder/prisopgaven.ashx

Prideaux, B., King, B., Dwyer, L., & Hobson, P. (2006). The hidden costs of cheap group tours—A case study of business practices in Australia. In J. S. Chen (Ed.), *Advances in hospitality and leisure* (pp. 51–71). Bingley: Emerald Group Publishing Limited.

Reece, W. S. 2009. *The economics of tourism.* Upper Saddle River, NJ: Prentice Hall.

Rugg, D. (1973). The choice of journey destination: A theoretical and empirical analysis. *The Review of Economics and Statistics*, *55*(1), 64–72. doi:10.2307/1927995

Ryan, C., Minghui, S., & Xiaoyu, Z. (2013). Chinese tourists and souvenir shopping in New Zealand. *The China-New Zealand Tourism Research Centre Department of Tourism and Hospitality Management The University of Waikato Management School.*

Ryan, C., & Mo, X. (2002). Chinese visitors to New Zealand-Demographics and perceptions. *Journal of Vacation Marketing*, *8*(1), 13–27. doi:10.1177/135676670200800103

Schwartz, Z. (2007). Revenues and asymmetric information: How uncertainty about service quality and capacity management affect optimal advanced booking pricing. *Journal of Quality Assurance in Hospitality & Tourism*, *7*(4), 1–22. doi:10.1300/j162v07n04_01

Soman, D., & Gourville, J. T. (2001). Transaction decoupling: How price bundling affects the decision to consume. *Journal of Marketing Research*, *38*(1), 30–44. doi:10.1509/jmkr.38.1.30.18828

Ting, D. H. (2004). Service quality and satisfaction perceptions: Curvilinear and interaction effect. *International Journal of Bank Marketing*, *22*(6), 407–420. doi:10.1108/02652320410559330

Tourism New Zealand. (2016). *Three words: Approved destination status.* Retrieved March 28, 2019, from https://www.tourismnewzealand.com/news/three-words-approved-destination-status/

Tourism New Zealand. (2018). *China market overview.* Retrieved August 20, 2019, from https://www.tourismnewzealand.com/markets-stats/markets/china/

Velázquez, B. M., Saura, I. G., & Molina, M. E. R. (2011). Conceptualizing and measuring loyalty: Towards a conceptual model of tourist loyalty antecedents. *Journal of Vacation Marketing*, *17*(1), 65–81. doi:10.1177/1356766710391450

Wang, K.-C., Hsieh, A.-T., Chou, S.-H., & Lin, Y.-S. (2007). GPTCCC: An instrument for measuring group package tour service. *Tourism Management*, *28*(2), 361–376. doi:10.1016/j.tourman.2006.04.017

Wang, K.-C., Hsieh, A.-T., & Huan, T.-C. (2000). Critical service features in group package tour: An exploratory research. *Tourism Management*, *21*(2), 177–189. doi:10.1016/s0261-5177(99)00047-3

Wang, Y., & Davidson, M. C. G. (2010). Chinese holiday makers' expenditure: Implications for marketing and management. *Journal of Hospitality Marketing & Management*, *19*(4), 373–396. doi:10.1080/19368621003667101

Wang, Y., & Sheldon, P. J. (1996). The sleeping dragon awakes: The outbound Chinese travel market. *Journal of Travel & Tourism Marketing*, *4*(4), 41–54. doi:10.1300/j073v04n04_03

Wang, Y., Weaver, D. B., & Kwek, A. (2015). Beyond the mass tourism stereotype: Power and empowerment in Chinese tour packages. *Journal of Travel Research*, *55*(6), 724–737. doi:10.1177/0047287515592971

Wong, C. U. I., & McKercher, B. (2012). Day tour itineraries: Searching for the balance between commercial needs and experiential desires. *Tourism Management*, *33*(6), 1360–1372. doi:10.1016/j.tourman.2011.12.019

Xiang, Z., & Gretzel, U. (2010). Role of social media in online travel information search. *Tourism Management*, *31*(2), 179–188. doi:10.1016/j.tourman.2009.02.016

Yang, S., Isa, S. M., Wu, H., Ramayah, T., & Jermsittiparsert, K. (2020). Examining the role of destination image, self-congruity and trip purpose in predicting post-travel intention: The case of Chinese tourists in New Zealand. *Revista Argentina de Clinica Psicologica*, *29*(5), 1504–1517.

Yang, S., Isa, S. M. M., Ramayah, T., Wen, J., & Goh, E. (2021). Developing an extended model of self-congruity to predict Chinese tourists' revisit intentions to New Zealand: The moderating role of gender. *Asia Pacific Journal of Marketing and Logistics*, ahead-of-print. doi:10.1108/APJML-05-2021-0346

Yu, P. (2012). *The Ling-Tuan-Fei Tour to New Zealand: Poison, panacea or palliative?*. [Master dissertation, Auckland University of Technology]. Auckland University of Technology Library. http://orapp.aut.ac.nz/handle/10292/4728

Yu, X., Weiler, B., & Ham, S. (2002). Intercultural communication and mediation: A framework for analysing the intercultural competence of Chinese tour guides. *Journal of Vacation Marketing*, *8*(1), 75–87. doi:10.1177/135676670200800108

Zhang, Y., & Murphy, P. (2009). Supply-chain considerations in marketing underdeveloped regional destinations: A case study of Chinese tourism to the Goldfields region of Victoria. *Tourism Management*, *30*(2), 278–287. doi:10.1016/j.tourman.2008.07.004

Zhao, C. (2006). *New Zealand's destination image and the Chinese outbound market: A comparative study between the Beijing (North) and Guangdong (South) markets* [Master dissertation, Auckland University of Technology]. Auckland University of Technology Library. http://orapp.aut.ac.nz/handle/10292/137

10 Chinese tourists' behavior on package tours in Turkey

Vedat Acar

Introduction

China has a significant share of the world tourism market. For example, a United Nations World Tourism Organization (UNWTO) (2018) report showed that China was the top spender globally in 2017 at US$ 257.7 billion. The UNWTO (2019) report showed that approximately 10% of Chinese people had participated in international tourism, and it remained the world's largest spender at US$ 277 billion. The high spending tendency of Chinese tourists undoubtedly grabbed the attention of many countries. Turkey, however, has not benefited as much from this tourist market as other countries mainly because certain travel agents have monopolized and dominated its market. They do so by underselling tour packages (including flights) for about US$550–650 to Chinese tourists before trying to recoup the loss by charging high commissions and head taxes from carpet, leather, jewelry, and other shops. Because of the high commissions and head taxes, these shops have to sell their products at exorbitant prices (Acar et al., 2020). This, in turn, affects tourists' behaviors and damages their attitudes toward Turkey in the long run.

To dominate a tourism market, it is also necessary to understand the culture and the behaviors of target tourists. Compared to Western countries, Far Eastern tourists clearly have various distinct behavioral patterns from food and beverage preferences to social interactions. These cultural differences also affect their behaviors while on package tours. Although it is essential to understand such behavior patterns to benefit from this market, very few studies have investigated Chinese tourists participating in package tours in Turkey. Accordingly, the study reported in this chapter investigated six typical tourist behaviors (interaction, shopping, bargaining, tipping, food, and beverage preferences) of Chinese tourists in package tours in Turkey.

Literature review

To understand Chinese behaviors on package tours, it is important to review package tours and the travel agents serving them. In Turkey, package tours can be classified as séjour and itinerant package tours. Séjour means

DOI: 10.4324/9781003121329-12

"stay, trip" in English (Cambridge Dictionary, 2021), whereas itinerant refers to "travelling from place to place" (OWD, 2003). Those tourists who take part in séjour package tours are in search of rest and relaxation, while the tourists who partake in itinerant package tours seek for novelty and getting to know culture. Examples of séjour package tourists include visitors from countries like Russia, Germany, the Netherlands, Denmark, Belgium, Sweden, and Norway, who mainly buy transportation and accommodation services from travel agents and participate in daily tours. Examples of itinerant tourists include visitors from countries like China, Thailand, Taiwan, Japan, Malaysia, South Korea, and Singapore who join tours to destinations like Istanbul, Çanakkale, Bergama, Sardes, Ephesus, Pamukkale, Antalya, Konya, and Nevşehir (Cappodocia).

The findings of various studies show that most Japanese, South Korean, and Chinese tourists travel with their families and book through travel agents (e.g., İbiş & Batman, 2018a, 2018b). According to Navarro (2016), Chinese tourists prefer organized group tours because of the language barrier, visa difficulties, fear of adventure, a collective community structure, and price benefits. A proper analysis of the package tour behaviors of Chinese tourists depends on a good understanding of the travel agents that serve this market. As mentioned above, some Turkish travel agents have tried to monopolize the Far Eastern market. However, this poses a significant risk both to competing travel agents and tourist shops selling products like carpets, leather, and jewelry. Acar et al. (2020) highlighted the monopolistic role of such travel agencies offering supposedly cheap tour packages. These practices entail cheap labor, a substandard hotel room, poor-quality dining services, and very high product prices. This, in turn, results in unsatisfied customers.

For both daily and package tours, travel agents in Turkey must appoint a licensed tour guide (Association of Turkish Travel Agencies [Türkiye Seyahat Acentaları Birliği] [TÜRSAB], 1972). The fees of tour guides are determined by The Republic of Turkey Ministry of Culture and Tourism under the Tourist Guidance Professional Law (2012) No. 6326. Although it is illegal to pay less, travel agents force tour guides to work for less because they sold the tour at too low a price. Furthermore, these agencies pressure tourist shops, restaurants, and hotels. For example, they force tourist shops to pay commission for the shoppers, hotels to sell rooms for much less than the door price, and restaurants to offer cheap meals.

Interaction

Like many other Asian tourists, one of the major problems Chinese tourists face is the language barrier. For example, Parasakul (2020) showed that more than half of Chinese tourists visiting Bangkok complain about language communication problems and suggested that salespeople should speak Chinese to increase the satisfaction of Chinese tourists. According to Li et al. (2011), Chinese tourists prefer group tours to avoid such cultural and language barriers, while Laitinen (2004) found that Chinese tourists do not enjoy free time on their own because of their cultural background and lack of English. Furthermore, based on their experiences in standard hotels in China, Chinese tourists expect free services like toothpaste, toothbrushes, combs, shampoos, lotions, slippers, and shoe mitts (Li et al., 2011). However, because of the language barriers while abroad, they cannot request such items from hotel staff.

The tour programs that Chinese tourists join tend to be very tightly scheduled (Li et al., 2011; Melubo & Kisasembe, 2020), which prevents them from interacting with and getting to know other tourists and local communities. Similarly, in Turkey, Chinese tourists usually visit destinations like İstanbul, Çanakkale, İzmir, Denizli, Antalya, Konya, Nevşehir, and Ankara by tour bus in just one week, which leaves no time for interacting with other tourists and local communities.

Shopping

According to Navarro (2016), Chinese tourists are notably fond of shopping, with those traveling abroad tend to buy souvenirs for their relatives and friends back home. For example, Chinese tourists visiting Singapore prefer to buy gemstones (Navarro, 2016) while those visiting the United States prefer antique watches, jewelry, health products, and running shoes (Xu & McGehee, 2012). Li and Ryan (2018) found that the gift-giving tradition was an important shopping motivation for Chinese visitors to North Korea, with traditional Chinese medicine being an important shopping item. In Hawaii, Chinese tourists prefer discounted products, with a significant difference between first-time visitors and others. For example, first-time visitors tended to buy traditional Hawaiian gifts and Hawaiian chocolate, whereas those who had visited Hawaii ten times or more preferred alcohol (Agrusa et al., 2011). Finally, Chinese shoppers in Metropolis, Bangkok, most preferred food products, cosmetics and medicine, clothing and accessories, clothes, and artworks and souvenirs (Parasakul, 2020). It should also be noted that, while outbound Chinese tourists enjoy shopping, forced shopping has a negative effect on them, so service providers in Western tourist destinations must become familiar with Chinese lifestyles, particularly their food culture (Li et al., 2011).

Currently, the shopping tendencies of Chinese tourists visiting Turkey are mostly determined by bilateral agreements between travel agents and tourist shops. During a week-long package tour, Chinese tourists visit a wide range of tourist shops selling items like leather, carpets, textiles, jewelry, and Turkish delight. These tourists also like to go on the extra tours offered (balloon trips, jeep safaris, Turkish nights, etc.). However, to get them to spend more money, it is necessary to employ Chinese-speaking staff (Xu & McGehee, 2012). In Turkey, sales staff working in leather, jewelry, and carpet shops are very persuasive and have good command of tourists' languages and cultures. Therefore, compared to its rivals, Turkey already has an advantage.

Food and beverage preferences

While transportation, accommodation, and dining are all essential components of a package tour, particular attention should be paid to food and beverages, especially for Far Eastern tourists. Chinese tourists can be dissatisfied by a lack of food and beverage services that appeal to their tastes. For example, Finnish tour guides reported that Chinese tourists do not like Finnish food containing a lot of dairy products due to their different eating and drinking habits (Laitinen, 2004).

Despite these issues, Hong Kong Chinese tourists (Wong & Lau, 2001) and Mainland Chinese tourists (Chang et al., 2010) try local food and beverages during their trips. However, Agrusa et al. (2011) found that Chinese tourists prefer ethnic Chinese to local foods while visiting Hawaii. Similarly, Melubo and Kisasembe (2020) confirmed that, apart from some adventurous individuals willing to experience African flavors, many Chinese tourists in Tanzania want flavors to appeal their own palates. Li et al. (2011) report that Chinese tourists complain about the Western-style foods that are too sweet, fried, unhealthy, or include too many uncooked or cold dishes (including iced water). In contrast, Chinese cuisine includes a wide variety of vegetables, very little or no milk, and salty foods opposite sugary foods. They conclude that the way to win over Chinese tourists is to know their tastes.

According to Navarro (2016), eating out is a tradition for Chinese people, so quality concerns in restaurants are a particular concern on tours. Indeed, according to the Finnish tour guides, the taste of the food served to Chinese tourists in Finland was far from the tastes of dishes served in restaurants in China (Laitinen, 2004).

A Turkish guide who lived in South Korea for seven years noted that Chinese tourists bring instant noodles that can be eaten simply by adding hot water, so they require kettles in their hotel rooms (Acar, 2018). Chinese tourists on package tours in Turkey usually eat hotel buffet breakfasts and dinners, while lunches depend on the destination visited and tour operation. However, very cheap package tours offer very poor food service. On the other hand, Far Eastern tourists can sometimes find their traditional dishes during package tours in Turkey.

Bargaining

The Turkish Language Institution dictionary defines bargaining as a meeting between the parties during a shopping process to get them to accept the most favorable price for them (Türk Dil Kurumu, 2021). Lee (2000) revealed that Chinese tourists have a more competitive bargaining style than

Americans (Lee, 2000). Although cultures differ in bargaining behaviors, pricing policies in the countries visited also determine whether tourists bargain or not. For example, British tourists attempt to bargain in Turkey, Tunisia, Morocco, India, and Egypt (Kozak, 2016). During an observation study in Cappadocia in 2016, a Chinese tourist in a tour group asked the shop assistant in a local dried-fruit shop whether he could get any discounts and reacted with joy when his request was accepted (Acar, 2018). Similarly, during an observational study in İstanbul during 2016, the tour guide warned Pakistani tourists to request at least 30% discount when buying anything at the Grand Bazaar (Acar & Tanrısevdi, 2019a).

A Japanese-speaking tour guide with 16 years' experience noted how Japanese package tourists were unaccustomed to bargaining. Nevertheless, the tour guides always encouraged them to request discounts while in Turkey (Acar, 2018). Similarly, South Korean tourists in Cappadocia tended to bargain (Tanrısevdi & Akyurt Kurnaz, 2018), which supports the findings of other studies carried out in Turkey.

In short, tourists visiting Turkey try to bargain, regardless of their nationality or country. This behavior is explained by two factors. The first is the absence of standard price labeling, for example, in carpet and leather shops (Acar et al., 2020). As one German-Dutch-English =speaking guide noted:

> Now the biggest problem in Turkey is that, have a look, there's no label for anything at the stores. You can never see this in any European countries. Otherwise, the store officials will be fined directly. The tourist looks at the price of the sweater but finds no price label on it.
>
> (Acar, 2018, p. 264)

The other factor is extreme discounts. According to research conducted in İstanbul, Cappadocia, Antalya, and Kuşadası between July and October 2018, extreme price reductions from the labeled price were a highlight of the tourists' visits (Acar, 2018).

Regardless of their country of origin, tourists see bargaining as an essential characteristic of shopping in Turkey. Therefore, comparisons of bargaining behavior based on the tourist's nationality or culture may be invalid as the main reason they bargain in Turkey a fear of being cheated.

Tipping

A tip is a symbolic indicator of service satisfaction. However, the meaning of this behavior varies across cultures. For example, Americans traditionally leave a 15–20% tip for restaurant staff (Lynn, 2010), whereas European restaurants include a similar gratuity in the bill (Reisinger, 2009). Comparatively, Americans tip based on the service they receive, whereas British and Australians are less likely to tip while Mainland Chinese tip the least (Dewald, 2001). Tips also have a very different meaning in Far Eastern cultures.

Specifically, there is a borderline between gift-giving and bribery in countries like Japan and China, so tipping may be considered as a form of bribery there (Hofstede et al., 2010).

Nevertheless, although tipping is not part of Chinese culture, restaurants and hotels have started adding a 10% service charge bills (Dewald, 2001). The service fee applied in many luxury hotels in China is not shared with the hotel staff, although one five-star hotel shared 1%of the service charge with the staff. In China, hotel employees do not expect or solicit tips, so tips mainly come from tourists rather than local people, who tend not to tip (Dewald, 2021). Regarding Chinese visitors to the United States, the average tip in restaurants is 5.30%, which is well below 15–20% (Lee & Dewald, 2016).

According to a Turkish tour guide who had lived in South Korea for seven years, the tip is included as a service fee in bills in Korea but not in China. He also remarked that some people were surprised and even insulted when he left a tip during his visit to China (Acar, 2018). For tipping in Turkey, daily and package tours should be considered separately. The tourists' nationalities are an important factor determining tipping in package tours, whereas tourists tip less in daily tours because they know that the tour will end in a very short time. Service providers in Turkey generally advise Far Eastern package tourists to leave tips.

Methodology

Tracy (2013) emphasizes that qualitative methods cover not only interviews and participant observation but also document analysis. Lincoln and Guba (1985) state that those methods such as interview, observations, document analysis, and unobtrusive clues are appropriate for humanly implemented inquiry. According to CDC (2018), the document review means collecting and reviewing the existing documents such as reports, program logs, performance ratings, funding proposals, meeting minutes, and marketing materials (CDC, 2018). From this viewpoint, the present study used document review to investigate the typical tourist behaviors of Chinese tourists in package tours in Turkey. The six areas of the literature reviewed above were derived from previous investigations of the behaviors of tourists in guided tours (Acar, 2018; Acar & Tanrisevdi, 2019a, 2019b, 2021).

Undoubtedly, tour guides have the most contact with tourists from the beginning to the end of a package tour. Accordingly, three tour guides with experience in Turkish package tours with Chinese participants evaluated their behaviors in terms of the six areas from the literature review. Thus, the study aimed to support the relevant literature.

The participants were selected from the author's colleagues and a professional licensed tour guide. The main inclusion criterion was experience with package tours that included Chinese tourists. In January 2021, each participant was invited through a phone conversation. The three participants, one living in Antalya and the others in Kuşadası, accepted that invitation. Then, they were asked to evaluate the Chinese tourists' behaviors in terms of interaction, bargaining, tipping, shopping, and food and beverage preferences. The participants' views about the Chinese tourists' behaviors were sent to the researcher via e-mail in February 2021.

The participants were referred to as Tour Guides 1, 2, and 3 in the Results section. To ensure the credibility of qualitative data, Lincoln and Guba (1985) recommend prolonged engagement, persistent observation, and triangulation. Thus, tour guides were included in this study because they have "prolonged engagement" and "persistent observation" throughout each package tour. Another important criterion for qualitative research is "transferability". To better understand the findings, everything that a reader may need to know must be specified on describing (Lincoln & Guba, 1985). In the present study, thick description was applied to achieve transferability, and the researcher referred to direct quotations to present the tour guides' views about the Chinese tourists' behaviors.

Results

Cultural and language barriers are key reasons why Chinese tourists prefer package tours (Laitinen, 2004; Li et al. 2011; Navarro, 2016; Parasakul, 2020). Because some tourists want to maintain their everyday habits even while visiting other countries, they feel more confident if they can remain in an environmental bubble reflecting their homeland throughout their trip (Doğan, 2004). They feel more comfortable if they can act together with tourists of the same nationality. Surprisingly, however, Chinese people score low on Hofstede's uncertainty avoidance dimension and do not mind ambiguity (www.hofstede-insights.com). Thus, it is difficult to explain why Chinese tourists should prefer the certainties of package tours.

One possible explanation is that Chinese tourists choose package tours to Turkey because of the cheap prices. As Tour Guide 1 noted, they could buy a one-week trip, including flight, hotels, tours, and lunches, for under US$1,000. This confirms a similar report from a leather shop sales assistant in Selçuk, İzmir that these package tours only cost US$580–650 in 2019 (Acar et al., 2020). However, even a one-way flight from China to Turkey costs more nowadays.

Previous research has shown that Chinese tourists enjoy shopping in various destinations (Agrusa et al., 2011; Li & Ryan, 2018; Navarro, 2016; Parasakul, 2020; Xu & McGehee, 2012), particularly to buy souvenirs for their friends and relatives (Li & Ryan, 2018; Navarro, 2016). Preferred products include antique watches, jewelry, health products, and running shoes (Xu & McGehee, 2012), alcoholic drinks, discounted products, and traditional gifts (Agrusa et al., 2011), food items, cosmetics and medicine, clothing and accessories, and artworks (Parasakul, 2020). In Turkey, Chinese tourists visit tourist shops selling carpets, leather goods, jewelry, pottery, gemstones, etc. According to Tour Guide 1, the most popular purchases are Turkish delight, rose water, and jewelry, while hot-air balloon trips, jeep safaris, and parachute jumps were the mostly preferred optional tours. Acar and Tanrisevdi (2021) also found that South Korean package tourists were very interested in buying similar optional tours while in Turkey. Thus, optional tours may be among the important purchases for Far Eastern tourists visiting Turkey. On the other hand, the tour guides also noted that travel agents tried to sell those optional tours to make a profit:

> Conglomerate X can organise tours at lower prices since many places such as restaurants, stores, hotels, and accommodation facilities in Anatolia belong to that conglomerate. Thus, it forces the guide and the tour leader to sell products to the customers. Most of the balloon companies in Cappadocia are also controlled by this conglomerate. This gives a great competitive advantage.
>
> [Tour Guide 1]

> The travel agent did not give any advance payments, just the phone numbers to buy museum tickets on the black market! Furthermore, the costs of the tour were also covered by the money collected from the customers for the extra tour sales. For example, customers who purchased a balloon tour in Cappadocia could not fly due to the weather conditions or the lack of space, but the money they paid was spent on tour costs. Finally, the customers asked for their money back… Phone traffic with the travel agent, the customers' revolt! On the last day, the officials of the travel agent returned the customers' money to them at the airport.
>
> [Tour Guide 3]

Chinese people have different eating and drinking habits compared to those in Western societies (Agrusa et al., 2011; Laitinen, 2004; Li et al., 2011; Melubo & Kisasembe, 2020). Tour Guide 1 and Tour Guide 2 stated that Chinese tourists wanted to eat familiar food during their visit. According to Tour Guide 2, the travel agents took their tourist groups to Chinese restaurants twice during the tours. This supports Acar and Tanrisevdi (2021), who found that South Korean tourists ate at Korean restaurants twice during a package tour. Tour Guide 1 noted that Chinese tourists generally prefer very spicy and dried foods, so Turkish food did not suit them, except for kebabs. Tour Guide 1 also stated that Chinese tourists complained about the food during their trip. Those who did not want to eat the offered lunch preferred to go to Chinese restaurants even if they had to pay an extra charge.

Regardless of their nationality, tourists visiting Turkey tend to bargain (Acar, 2018; Acar & Tanrisevdi, 2019b, 2021; Kozak, 2016). This matches the experiences of Tour Guide 1 and Tour Guide 2 with Chinese tourists:

> Before traveling to Turkey, Chinese tourists read the comments on the internet about the shops, and they usually come Turkey with a bad

> impression. For this reason, they mostly avoid shopping in the tourist shops that the travel agents make them visit. Even if they go shopping, they bargain. They ask for a much larger reduction than the discounts offered by the shop assistants. I've witnessed several times that they asked for a reduction of 50–60–70%. Nevertheless, those demands were accepted by the salespeople.
>
> [Tour Guide 1]

> Chinese tourists love to bargain. On one tour, I remember that after a Chinese tourist asked the carpet seller, 'What is the final price?', he received the answer, 'US$1100', and the tourist told him, 'Let's settle for US$150' … I once witnessed in Pamukkale a group of ten friends putting ten Brand X leather jackets, which cost 100 Euros each, into ten different baskets, coming to the check out and saying they would only pay 100 Euros in total for ten jackets which cost 1,000 Euros. The cashier laughed and rejected. Most of the cashiers told me they were suffered from this matter.
>
> [Tour Guide 2]

Finally, various studies have investigated the tipping habits of different tourist groups (Acar, 2018; Dewald, 2001, 2021; Hofstede et al., 2010; Lee & Dewald, 2016; Lynn, 2010; Reisinger, 2009). In particular, there are notable differences between Western and Eastern civilizations. Supporting this, Tour Guide 1 and Tour Guide 2 reported that Chinese package tour tourists are asked to pay predetermined tips in Turkey. Tour Guide 2 noted that Chinese tourists lack a tradition of tipping. Similarly, Acar and Tanrisevdi (2021) reported that South Korean package tour tourists pay predetermined tips, including to the tour guide, tour leader, and bus driver. Thus, tipping is a required one rather than optional for the Far Eastern package tour tourists in Turkey.

Conclusion

This chapter discussed the typical tourist behaviors of Chinese tourists participating in package tours in Turkey. Chinese tourists are well known as the world's largest spenders (UNWTO, 2018, 2019). However, Turkey appears not to get enough of this tourist market share. The main reasons are monopolistic travel agents and very cheap package tours are offered to Chinese tourists (Acar et al., 2020). To prevent this monopolization and allow others to benefit from this market share, the Turkish government and local administrations must implement and monitor a minimum base price for package tours.

Shop visits for leather, carpets, jewelry, pottery, textiles, etc. are essential components of package tours in Turkey. However, the lack of standard price labeling and discrepancies between the official price and lowest accepted price (Acar, 2018; Acar et al., 2020; Acar & Tanrisevdi, 2019b, 2021) undoubtedly encourage tourists to bargain, regardless of their culture or country. To change this image, local and central administrations must increase their supervision and impose sanctions against shops offering disproportionate prices.

Cappadocia is an important destination for Turkey's Far Eastern tourist market, with hot-air balloon tours being a prominent attraction. A few years ago, a consortium was formed to unite Cappadocia's balloon tour companies, which was then investigated by the Competition Authority, for violating competition rules (Airporthaber, 2019; Turizm Haber Merkezi, 2019). On March 16, 2021, evidence was requested from the 41 companies operating as hot-air balloon companies and tourism agents in Cappadocia, and the investigation was concluded on April 1, 2021. As a result, the consortium's joint reservation and sales activities were stopped (Turkish Competition Authority, 2021). Given that such tours are among the most popular optional tours for Far Eastern tourists, it was critical to prevent such monopolization attempts for the future of Turkish tourism. Finally, food and beverage services are a key area of dissatisfaction for Chinese tourists as they can only experience their own cuisine two or three times during a package tour. That is why Turkish tour operators should take Chinese tourists' food and beverage expectations into consideration while creating tour itinerary.

References

Acar, V. (2018). *Investigating international tourists' behaviors: A study on tourists taking part in guided tours in Turkey* [Unpublished Doctoral Dissertation]. Aydin Adnan Menderes University, Turkey.

Acar, V., Karaosmanoğlu, K., & Şenbahar, Y. (2020). Turistik halı ve deri mağazaları ile seyahat acentaları arasındaki güncel sorunları belirlemeye yönelik nitel bir araştırma: Kuşadası ve Selçuk örneği [A qualitative research on determining current issues between touristic carpet and leather shops and travel agencies: The case of Kusadasi and Selcuk]. *Marmara Üniversitesi Öneri Dergisi, 15*(54), 610–643. https://dergipark.org.tr/tr/download/article-file/979355

Acar, V., & Tanrisevdi, A. (2019a). Paket turlarda turist davranışlarının incelenmesi: İstanbul'u ziyaret eden Pakistanlı turistler üzerinde nitel bir araştırma [Investigation of tourist behaviours in packaged tours: A qualitative research on Pakistani tourists visiting Istanbul]. *Manas Sosyal Araştırmalar Dergisi, 8*(4), 3931–3957. https://dergipark.org.tr/tr/download/article-file/826792

Acar, V., & Tanrisevdi, A. (2019b). Participant observations on Iranian tourists taking part in guided excursions, the case of Kusadasi, Turkey. *Manas Journal of Social Studies, 8*(2), 2187–2213. https://dergipark.org.tr/tr/download/article-file/703120

Acar, V., & Tanrisevdi, A. (2021). Paket tur katılımcısı turistlerin davranışlarının incelenmesi: Güney Koreli turistlere yönelik bir katılımlı gözlem çalışması [Investigating the behaviors of package tour participants: A participant observation research on South Korean tourists]. *Marmara Üniversitesi Öneri Dergisi, 16*(56), 792–824. https://dergipark.org.tr/tr/download/article-file/1361106

Agrusa, J., Kim, S. S., & Wang, K.-C. (2011). Mainland Chinese tourists to Hawaii: Their characteristics and preferences. *Journal of Travel & Tourism Marketing, 28*(3), 261–278. doi:10.1080/10548408.2011.562853

Airporthaber. (2019, November 25). Sıcak hava balonu işletmelerine Rekabet Kurumu şoku! *Airporthaber.* Retrieved March 14, 2021, from https://www.airporthaber.com/havacilik-haberleri/sicak-hava-balonu-isletmelerine-rekabet-kurumu-soku.html

Associaton of Turkish Travel Agencies (Türkiye Seyahat Acentaları Birliği). (1972). *Travel Agents and Association of Travel Agents Act.* Retrieved February 7, 2021, from https://www.tursab.org.tr/dosya/1063/1618sayiliyasaingilizce2013-3_1063_5073648.pdf

Cambridge Dictionary. (2021). Séjour. Retrieved December 12, 2021, from https://dictionary.cambridge.org/dictionary/french-english/sejour

Centers for Disease Control and Prevention (CDC). (2018, November 19). Data collection methods for evaluation: Document review. Retrieved December 19, 2021, from https://www.cdc.gov/healthyyouth/evaluation/pdf/brief18.pdf

Chang, R. C. Y., Kivela, J., & Mak, A. H. N. (2010). Food preferences of Chinese tourists. *Annals of Tourism Research, 37*(4), 989–1011. doi:10.1016/j.annals.2010.03.007

Dewald, B. (2021). Service charges and tipping: A case study of the Chinese hospitality industry. *Journal of Tourism and Sports Management, 3*(2), 307–314.

Dewald, B. W. A. (2001). Tipping by foreign tourists. *Pacific Tourism Review, 5*(1–2), 43–50.

Doğan, H. Z. (2004). *Turizmin sosyo-kültürel temelleri [Socio-cultural foundations of tourism].* Ankara: Detay Yayıncılık.

Hofstede, G., Hofstede, G. J., & Minkov, M. (2010). *Cultures and organizations: Software of the mind.* New York: McGraw Hill.

İbiş, S., & Batman, O. (2018a). Japonya ve Güney Kore'den İstanbul'a gelen turistlerin seyahat motivasyonları ve yeniden ziyaret etme niyeti üzerine bir araştırma [A study on travel motivations and revisit intention to Istanbul from Japan and South Korea tourists], *Seyahat ve Otel İşletmeciliği Dergisi, 15*(2), 246–262. https://dergipark.org.tr/tr/download/article-file/523853

İbiş, S., & Batman, O. (2018b). Türkiye'ye gelen Çinli turistlerin seyahat motivasyonları üzerine bir araştırma, *Hitit Üniversitesi Sosyal Bilimler Enstitüsü Dergisi, 11*(1), 455–476. doi: 10.17218/hititsosbil.346388

Kozak, M. (2016). Bargaining behavior and the shopping experiences of British tourists on vacation. *Journal of Travel & Tourism Marketing, 33*(3), 313–325. doi:10.1080/10548408.2015.1051199

Laitinen, K. (2004). *The perceptions of tour guides of Finland as a travel destination for Chinese tourists* [Unpublished master dissertation]. Bournemouth University.

Lee, D. Y. (2000). Retail bargaining behaviour of American and Chinese customers. *European Journal of Marketing, 34*(1/2), 190–206. doi:10.1108/03090560010306287

Lee, P. C., & Dewald, B. (2016). Tipping practices of Chinese tourists in U.S. restaurants: An exploratory study. *Journal of Tourism and Hospitality Management, 4*(2), 17–34. doi:10.15640/jthm.v4n2a2

Li, F., & Ryan, C. (2018). Souvenir shopping experiences: A case study of Chinese tourists in North Korea. *Tourism Management, 64*, 142–153. doi:10.1016/j.tourman.2017.08.006

Li, X., Lai, C., Harrill, R., Kline, S., & Wang, L. (2011). When east meets west: An exploratory study on Chinese outbound tourists' travel expectations. *Tourism Management*, *32*(4), 741–749. doi:10.1016/j.tourman.2010.06.009

Lincoln, Y. S., & Guba, E. G. (1985). *Naturalistic Inquiry*. Beverly Hills, CA: Sage.

Lynn, W. M. M. (2010). Tipping. In A. Pizam (Ed.). *International encyclopedia of hospitality management* (pp. 651–652). Oxford: Butterworth Heinemann.

Melubo, K., & Kisasembe, R. (2020). We need Chinese tourists, but are we ready? Insights from the Tanzanian safari industry. *Journal of China Tourism Research*. doi:10.1080/19388160.2020.1811822

Navarro, O. (2016). Chinese and tourism industry of Singapore. *Journal of Administrative Management, Education and Training (JAMET)*, *12*(4), 137–143.

OWD. (2003). *Oxford wordpower: Dictionary for learners of English*, M. Steel (Ed.). Oxford: Oxford University Press.

Parasakul, L. (2020). Assessing Chinese Tourists' satisfaction with their shopping experiences in Bangkok Metropolis. *Journal of China Tourism Research*, *16*(2), 279–296. doi:10.1080/19388160.2019.1638864

Reisinger, Y. (2009). *International tourism: Cultures and behaviours*. Oxford: Butterworth-Heinemann.

Tanrısevdi, A., & Akyurt Kurnaz, H. (2018). Hediyelik eşya satın alma davranışında kültürel özellikler: Hediyelik eşya satıcıları açısından bir inceleme [Cultural features in souvenir purchasing behavior: A review of souvenir shop sellers]. *Journal of Travel and Tourism Research*, *13*, 1–18. https://dergipark.org.tr/tr/download/article-file/848418

Tourist Guidance Professional Law. (2012). Retrieved March 17, 2021, from https://www.resmigazete.gov.tr/eskiler/2012/06/20120622-2.htm

Tracy, S. J. (2013). *Qualitative research methods: Collecting evidence, crafting analysis, communicating impact*. Chichester: Wiley-Blackwell.

Turizm Haber Merkezi. (2019, November 23). Dorak Holding'e Kapadokya'da balon şoku. *Turizmhabermerkezi.net*. Retrieved March 14, 2021, from https://www.turizmhabermerkezi.net/dorak-holding-e-kapadokya-da-balon-soku/11846/

Türk Dil Kurumu. (2021). Pazarlık. Retrieved March 17, 2021, from https://sozluk.gov.tr/

Turkish Competition Authority. (2021). *Kapadokya bölgesinde sıcka hava balonu işletmesi ve turizm acentesi oarka faaliyet gösteren 41 teşebbüs hakkında yürütülen soruşturma sonuçlandı*. Retrieved April 30, 2021, from https://www.rekabet.gov.tr/tr/Guncel/kapadokya-bolgesinde-sicak-hava-balonu-i-8789c99bb992eb11812d00505694b4c6

United Nations World Tourism Organization. (2018). *UNWTO tourism highlights 2018 edition*. https://www.e-unwto.org/doi/epdf/10.18111/9789284419876

United Nations World Tourism Organization. (2019). *International tourism highlights 2019 edition*. https://www.e-unwto.org/doi/pdf/10.18111/9789284421152

Wong, S., & Lau, E. (2001). Understanding the behavior of Hong Kong Chinese tourists on group tour packages. *Journal of Travel Research*, *40*(1), 57–67. doi:10.1177/004728750104000108

www.hofstede-insights.com. *Country comparison*. Retrieved November 1, 2021 from https://www.hofstede-insights.com/country-comparison/china/

Xu, Y., & McGehee, N. G. (2012). Shopping behavior of Chinese tourists visiting the United States: Letting the shoppers do the talking. *Tourism Management*, *33*(2), 427–430. doi:10.1016/j.tourman.2011.05.003

11 Destination food image

A systematic literature review to assess future perspectives

Fangli Hu and Han Shen

Introduction

Food tourism, a kind of special interest tourism, is gradually becoming a new trend and new direction for future tourism development. More and more tourists' destination decisions are influenced by food. According to the survey data from *Travel Weekly* in 2018, 64% of tourists chose destinations based on the attractiveness of local cuisine.[1] Booking.com, an online booking platform for hotels and accommodations, also found that among Chinese tourists, up to 75% of groups believe that tasting local food is an extremely important part of their journey, and even the primary determinant.[2] Thus, as an indispensable part of the global intangible cultural heritage and tourism elements, "food" has gone beyond the most basic function of satiation and begun to meet tourists' more advanced pursuit and enjoyment. Food is playing an increasingly important key role in their destination decision-making, and food tourism motivation has become an important driving factor of tourism.

Moreover, with the development of mass tourism, tourists' preferences, activity methods, and activity content have changed greatly. The focus on food itself has gradually shifted to various activities such as ingredients, food and its processing, cooking process, related festivals, transactions, and training as well as the extension of the environment related to food such as countryside and agriculture. According to the World Food Travel Association's 2016 survey, 80% of leisure tourists will visit a destination because of a gourmet cooking event.[3] Accordingly, food tourism now requires more experience. In addition, it is worth noting that food and beverage expenditures account for up to one-third of the total global tourism expenditures (Meler & Cerovic', 2003). Consequently, food tourism has a fairly huge development potential.

In order to develop food tourism, governments around the world are racing to formulate relevant strategies and policies. France, Spain, Malaysia, Australia, South Korea, Japan, Brazil, Mexico, Egypt, and many other countries have all listed food tourism as a national tourism development strategy. Furthermore, the development of food tourism has also been recognized and supported by international authorities such as the United

DOI: 10.4324/9781003121329-13

Nations World Tourism Organization (UNWTO) and the United Nations Educational, Scientific and Cultural Organization (UNESCO) and advocated and promoted by private sector and non-governmental organizations such as the World Food Travel Association. Relevant policies of food tourism in some countries are shown in Table 11.1.

Table 11.1 Summary of relevant policies on food tourism in some countries

Country	*Relevant policies*
France	Using **food and wine** as the core product to closely integrate culture and tourism and focus on establishing **international food tourism brands**, including "the art of surprising you" in Alsace, and "the art and joy of living" in Burgundy and other typical cases
Spain	Determining food tourism as an important planning content of the national development strategy, with "**Tasting Spain**" as the planning and promotion theme, and using the combination of people, food and wine, and other characteristic elements to showcase local culture
Portugal	Focusing on the development of "food and wine tourism" and including it in the national tourism development strategy
Greece	According to the survey results of local tourism enterprise associations, **food tourism is seen as one of the key sources of local tourism core competitiveness**
Malaysia	Local food is rich in variety, including a variety of **ethnic food**. The core strategy is to promote the development of **Halal food** while focusing on the similarities and differences of cuisine
Singapore	Setting up a special award-**Hawker Food Award** as an incentive measure to protect the local food and cultural heritage and develop high-quality tourism
Thailand	Taking "**World Kitchen**" as the theme of food tourism development
Australia	It can be regarded as a pioneer in the development of **wine tourism** in the world. It explores the policies and measures for the development of wine tourism by holding related events – the first Global Wine Tourism Conference
New Zealand	Identifying "**Tasting New Zealand**" as the theme of food tourism development and including it in the three themes of New Zealand's national tourism development strategy "The Tourism New Zealand"
Ireland	Regarding "**Pure Ireland, Pure Taste**" as the theme of food tourism development
South Korea	Since 1999, the Korean government has been working to turn the country's traditional food into tourist attractions
	Regarding food tourism as a national key development project after the "Korean Wave" and vigorously promoting **the internationalization of Korean food**. In 2009, the Korean government formulated and passed the "Government Funding of Korean Food Foundation Act" and established a multi-sectoral joint Korean food national development agency, including the Ministry of Culture and Tourism, the Ministry of Foreign Affairs, and the Ministry of Agriculture, Forestry, and Fisheries.

(*Continued*)

Country	*Relevant policies*
Japan	Incorporating food tourism into the national policy of tourism. In 2006, the Japanese government established the "Basic Law for the Promotion of Tourism Establishment". The Ministry of Land, Infrastructure, Transport and Tourism **regards food tourism as an important part of enhancing tourism attractiveness and promoting regional revitalization**. The promotion of rural revitalization through tourism is not just about simple corporate profits, but aims to promote the sustainable development of tourism. To attract international tourists, the Ministry of Agriculture, Forestry, and Fisheries has also formulated the "**Savor Japan Program**" and designated five project test areas.

Forty years of development have witnessed the prosperity of China's tourism market, which has entered a stable period. Additionally, globalization further leads the great progress of China's outbound tourism market and intensifies the market competition among destinations. What's more, the behavior of tourists is becoming more and more rational. They can obtain rich travel information from various channels including mass media and social media to choose the most attractive and suitable tourism destination. Accordingly, increasingly, mature tourist behavior has also intensified market competition in destinations.

However, research shows that a unique destination image can help a destination stand out in the intense market competition, especially many destinations with rich food resources. Food has been widely recognized as a pretty important element in the destination product portfolio and can be used as a primary or secondary tourism motivation (Quan & Wang, 2004), thereby increasing the value of the destination image (Boniface, 2003; Kivela & Crotts, 2006; Long, 2004). Meanwhile, according to a survey by the China Tourism Academy and Ctrip, food will replace shopping as the top priority for Chinese outbound tourists. Where to eat local cuisine is the most important information for Chinese outbound tourists.[4] Therefore, based on the huge scale and strong spending power of Chinese outbound tourists, it is necessary to figure out the destination food image perceived by Chinese outbound tourists. Thus, this chapter aims to review the literature on destination food image, summarize its research status, provide some directions for its future research, hoping to make some theoretical contributions to relevant research, and also provide suggestions for the management and marketing of destination food image, especially for Chinese outbound tourists.

Methodology

A systematic literature review (SLR) of articles on destination food image is applied in this paper. This method is beneficial to systematically summarize the relevant literature published before and more comprehensively grasp the research progress of the destination food image. As a result, this chapter uses the items "food tourism", "culinary/cuisine tourism", "gastronomy/gastronomic/gourmet tourism", "tasting tourism", and "food image" to search for related articles from the five databases of ScienceDirect, Emerald, Web of Science, Google Scholar, and Scopus. A total of 131 related articles in the period 1980–2021 were retrieved. Through careful study of these literatures, this chapter systematically clarifies the current research status of destination food image, including definition, measurement, and its relationships with other variables, and looks into the future research directions, in order to provide reference for deepening and expanding the theoretical research of destination food image. It is also expected to provide guidance for the practice of improving the destination food image.

Results

Food tourism started in the 1980s and has developed rapidly in recent years. Since the World Tourism Organization released a global report on the development trend of world food tourism in 2012, scholars have begun to focus on the development of food tourism and start research on food tourism. There is currently no uniform definition of food tourism, and many scholars define food tourism based on different emphases, including food tourism, culinary tourism, cuisine tourism, gastronomy/gastronomic tourism, gourmet tourism, and tasting tourism. Food tourism refers to the focus on ingredients, food and its processing, cooking process, related events, trade, and training activities as well as environment related to food such as countryside and agriculture, not just the food itself. Cuisine tourism emphasizes food preparation and cooking process.

Culinary tourism has a broader meaning, including not only the style of food, but also the social environment of cooking (Ignatov & Smith, 2006). Gastronomy/gastronomic tourism focuses on exploration, experience, and enjoyment (Horng & Tsai, 2010). Gourmet tourism specifically refers to the tourist group of gourmets (Cho, 2019). Tasting tourism highlights tourists' feelings when tasting food. Among them, in culinary tourism, tourists have relatively mild and moderate interest in food and are more likely to be interested in other tourism resources of the destination such as local culture and natural landscape. They mainly learn about local lifestyle and highlight their own uniqueness by participating in food-related activities. In gastronomy/gastronomic tourism, cuisine tourism, and gourmet tourism, tourists tend to have a very strong interest in food. Food is the main motivation for them to travel (Hall & Sharples, 2003).

It should be noted that, unlike general food consumption, food tourism offers many possibilities for destinations to make full use of local culture by developing stories related to food (Ignatov & Smith, 2006). Moreover, tourists can also obtain the most authentic experience and sense of place through food tourism (Sims, 2009) (Table 11.2).

Table 11.2 Definitions of food tourism

Author (Year)	*Definitions*
Boniface (2003)	**Tasting tourism** Appreciation and taste of food and wine
Ignatov and Smith (2006)	**Culinary tourism** Buying or consuming local foods (including beverages) or observing and studying food production (from farming to cooking schools) during travel is a critical motivation or activity that forms part of cultural tourism
Tsai and Wang (2017)	**Food tourism** Tourists engage in food-related activities during their travels such as buying and enjoying local food
Santich (2004)	**Gastronomy/gastronomic** A tourist activity triggered purely by an interest in local cuisine

Current research on food tourism

In 2000, the first international conference on "food and Tourism" was successfully held in Cyprus. Since then, food tourism research began to develop slowly. However, due to the relatively late start of food tourism research, there are still few relevant articles. Through combing past relevant literature, this study finds that current research on food tourism mainly focuses on the following aspects.

First, the definition of food tourism. Before carrying out the in-depth research on food tourism, scholars coincidentally define its definition first. However, presently, there are many and miscellaneous definitions of food tourism, and no unified definition has been formed. Scholars generally divide the concept and categories of food tourism in a finer manner and define it according to these categories. Thus, there is a certain controversy about the definition of food tourism. Second, food tourists and their behavior. The core of destinations' effective management and marketing strategies is to fully understand their target audience. For food tourists, current scholars mainly discuss the classification of their types and the influencing factors of tourism behavior.

Tourist segmentation is the key to the correct management of destinations and helps to develop specific tourism products to meet the diversified needs of each market segment (Dann, 1977). In terms of the classification of food tourists, the current main criteria include the importance of food in destination decision-making processes, the degree of interest in food, food-related lifestyle, and behavior. The detailed content is shown in Table 11.3.

Understanding tourism motivation is the first step to analyze tourist behavior. On the basis of tourism motivation, scholars have defined the food tourism motivation and studied its measurement items. In addition, the discussion of food tourist behavior is mainly to analyze the factors that may

Table 11.3 Types of food tourists

Criteria	*Author (year)*	*Types*
Degree of interest in food	Hall and Sharples (2003)	High interest, moderate interest, low interest, and no interest
	Björk and Kauppinen-Räisänen (2016) and Nicoletti et al. (2019)	Experiencer, enjoyer, and survivor
Food-related lifestyle and behavior	Hjalager (2003)	Existential, experimental, diversionary, and recreation travelers
The importance of food and gastronomy in their destination decision-making processes	Boyne et al. (2003)	Types I, II, III, and IV. The importance of food is gradually decreasing
Expectations of gastronomy, importance of gastronomy and other four factors	Kivela and Crotts (2005)	Typical leisure visitors and gastronomy tourist
Their stated participation in food-related travel experiences	Robinson et al. (2018)	Four segments

affect it. Scholars have tried to introduce a variety of variables in the context of food tourism through empirical analysis to test their impact on tourists' behavioral intention.

At present, the antecedents that have been proven to affect the behavioral intention of food tourists include food brand equity (Horng et al., 2012), destination environment (Mak et al., 2012), experimental value (Tsai & Wang, 2017), local food consumption value and tourist attitudes (Choe & Kim, 2018), destination food image (Lai et al., 2019), culinary tourist motivation (Agyeiwaah et al., 2019), perceived authenticity and culture contrast (Antón et al., 2019), and memorable food experience and food involvement (Di-Clemente et al., 2020). Among them, destination food image, memorable food experience, food involvement, etc. (Levitt et al., 2019) also play a mediating role. Moreover, Ellis et al. (2018) found that demographic factors, such as gender, age and educational background, have a significant impact on the food tourist behavior. It is worth noting that in addition to exploring the moderating effect of tourists' general personality characteristics, such as destination familiarity (Horng et al., 2012), some scholars have also applied tourists' food-related personality traits to food tourism. For example, Lai et al. (2019) found that food neophobia significantly moderated the formation of destination food image (Table 11.4).

Third, food tourism marketing and management strategy. Aware of the key role of food tourism to the destination, scholars have tried to make conducive suggestions for its development. Thus, the research on the

Table 11.4 Measurement items of food tourism motivation

Author (year)	*Measurement items*
Fields (2002)	Four dimensions: physical motivation, cultural motivation, interpersonal relationship, social status, and prestige
Çela et al. (2007)	Three dimensions (12 items): motivations to support, taste and purchase, motivations for attending the festival, and motivations to support and learn about local food
McKercher et al. (2008)	Two dimensions (eight items): general tourism motivation factors and specific food tourism motivation factors
Kim et al. (2009)	Nine dimensions (20 items): exciting experience, escape from routine, health concern, learning knowledge, authentic experience, togetherness, prestige, sensory appeal, and physical environment
Smith et al. (2010)	Six dimensions (51 items): food event, event novelty, socialization, food product, support services, and essential services
Kim and Eves (2012)	Five dimensions (24 items): cultural experience, excitement, interpersonal relation, sensory appeal, and health concern
Su et al. (2020)	Six dimensions (26 items): taste of food, socialization, cultural experiences, core food-tourism appeals, traditional food appeals, and local destination appeals

development strategy of food tourism occupies a large proportion of food tourism research. Horng et al. (2013), on the basis of the theory of planned behavior, took Macao Food Festival and Taiwan cooking exhibition as examples and proposed to subdivide tourists to the food festival and carry out differentiated management and marketing. Hillel et al. (2013), taking the Negev as an example, believe that tapping unique local food resources and combining them with geographical and cultural factors can make the destination more attractive. Chaigasem and Tunming (2019) conducted a SWOT analysis of the internal and external environment of food tourism development in Qinghan District, Rojiang Province, Thailand, and proposed that food tourism development should focus on food experience, combine local cultural heritage, and mobilize surrounding forces for public cooperation, so as to achieve sustainable development of food tourism.

Destination food image

As a research hotspot in the tourism field, destination image refers to "an attitudinal concept consisting of the sum of beliefs, ideas and impressions that a tourist holds of a destination" (Crompton, 1979, p. 18), which plays a critical role in the success of destination development and destination marketing (Tasci et al., 2007). Destination image includes the projected and perceived images. The projected image means "the purposive projection of a single message regarding the destination deliberately conveyed by various stakeholders of the destination through marketing channels". For these stakeholders, destination image is a key factor for destination positioning and promotion (Pike & Ryan, 2004), as well as a tourism resource, and even a manifestation of destination identity and personality (Lin et al., 2011). The perceived image refers to "the tourists' perception of a destination as a received promotional version or as experienced directly" (Hunter, 2012, p. 420). For them, destination image can greatly affect their destination decision-making process and behavior.

In previous literature, the destination food image was only regarded as a sub-component of the overall image of the destination (Lai et al., 2019). But, in recent years, scholars have gradually begun to use it as a core tourism product for special research (Karim & Chi, 2010; Kim et al., 2012; Peštek & Cinjarevic, 2014), most of which are quantitative empirical studies. From host's perspective, "food culture" is an expression of a destination's food identity, cultural heritage, and local lifestyle. From the perspective of tourists, destination food mainly meets their needs to experience local cuisine and culture through activities such as food tourism, food festivals, and cooking classes including both functional and symbolic values (Mohamed et al., 2021).

Destination food image as a multidimensional concept

Destination image has always been regarded as a multi-dimensional concept. The more classical classifications include cognitive-affective-conative model (Gartner, 1994) and cognitive-affective dual structure (Baloglu & McCleary, 1999). Like the variable of destination image, destination food image is also a multidimensional concept (Chi et al., 2013). However, many scholars currently do not realize this point. Generally, they use the single dimension of food cognitive image to represent the entire destination food image. For instance, Kim et al. (2012) used three dimensions of healthy menu, peculiarity of food culture, and variety/harmony to measure the destination food image.

Sanchez-canizares and Castillo-Canalejo (2015) summarized the destination food image of Spanish cities in the minds of tourists as traditional gastronomy, service and hospitality, quality of dishes, atmosphere establishments, facilities, prices, and innovative dishes. When Choe and Kim (2018) studied the destination food image of Hong Kong, they focused more on food culture and food itself. Promsivapallop and Kannaovakun (2019) recognized the importance of unique cultural experience for the Thai food image and included them in their measurement. Only a few scholars have realized the influence of emotional factors and taken the dimension of food affective image into consideration (Lai et al., 2017; Peštek & Cinjarevic, 2014; Seo & Yun, 2015).

Measurement of variables

Cognitive image refers to "the sum of beliefs and attitudes of an object leading to some internally accepted picture of its attributes" (Gartner, 1994, p. 193). It involves functional and practical attributes of destination image including intangible and tangible constituents of local cuisine (Lai et al., 2019). In extant studies, few scholars have specifically distinguished and explored the food cognitive image, and most of them treated single food cognitive image as destination food image in general. Many scholars have further refined the items on the basis of Lertputtarak (2012) and incorporated some service and cultural factors, including variety and harmony (Kim et al., 2012), food distinctiveness and accessibility (Chi et al., 2013), nutrition and health benefits of food (Peštek & Cinjarevic, 2014), atmosphere establishments (Sanchez-Cañizares & Castillo-Canalejo, 2015), food safety and quality, unique culinary arts (Seo & Yun, 2015), dining places/restaurants (Lai et al., 2017), and table manners as well as food taste and popularity (Yasami et al., 2020).

Although there are certain differences in these dimensions, they are mostly concentrated on the local food itself and dining places and rarely take into account the factors such as people and food activities. With this in mind, Lai et al. (2017) started from the dual perspectives of tourists and stakeholders and tried to clarify the dimensions of the food cognitive image. On the basis of the research review of destination food image, they took into account all-round factors such as people, geographical environment, and food activities and divided the food cognitive image into six dimensions, namely place and geographic environment, food and cuisine culture, food and people, food quality, dining places/ restaurants, and food activities, with a total of 39 items. The items and dimensions of food cognitive image are summarized in Table 11.5.

Affective image is related to "the motives one has for destination selection" (Gartner, 1994, p. 196) and beliefs and attitudes about the destination (Baloglu & McCleary, 1999). There are very few studies on the food affective image. Peštek and Cinjarevic (2014) put forward the affective dimension of destination food image when analyzing tourists' perception of Bosnian culture and used exciting and arousing to measure food affective image. Seo and Yun (2015) retained exciting and added items such as pleasant, fulfilled, enjoyable, and contented to measure the food affective image. Based on PAD model, planned behavior theory, and previous studies, Lai et al. (2019) integrated eight items, namely excited, pleasant, satisfied, aroused, enjoyable, fun, good, and relax to measure the food affective image of potential Chinese tourists to Australia (Table 11.6).

The conative image "is analogous to behavior" (Gartner, 1994, p. 196). It evolved from cognitive image and affective image (Chen & Phou, 2013). Existing studies on the dimension of food conative image mainly regard it as the tourists' behavioral intention, including revisit intention (Lertputtarak, 2012), visit intention (Karim & Chi, 2010; Kim et al., 2012), tourist satisfaction

Table 11.5 Items and dimensions of food cognitive image from tourists' perspective

Author (Year)	*Items and Dimensions of Food Cognitive Image*
Kim et al. (2012)	Three dimensions (12 items): healthy menu, peculiarity of food culture, variety, and harmony
Lertputtarak (2012)	Four dimensions (nine items): a good cultural experience, unique serving style, delicious, nourishing food, and exotic cooking methods
Chi et al. (2013)	Three dimensions (20 items): food distinctiveness and accessibility, food diversity and enjoyment, food quality, and presentation
Peštek and Cinjarevic (2014)	Three dimensions (11 items): food uniqueness and cultural heritage, food quality and price, nutrition, and health benefits of food
Seo and Yun (2015)	Five dimensions (23 items): food safety and quality, attractiveness of food, health benefits of food, food culture, and unique culinary arts
Sanchez-Cañizares and Castillo-Canalejo (2015)	Seven items: traditional gastronomy, service and hospitality, quality of dishes, atmosphere establishments, facilities, prices, and innovative dishes
Lai et al. (2017)	Six dimensions (39 items): place and geographic environment, food and cuisine culture, food and people, food quality, dining places/ restaurants, and food activities
Choe and Kim (2018)	Five items: rich food culture, diverse food, traditional food culture, delicious food, and unique food
Promsivapallop and Kannaovakun (2019)	Five dimensions (28 items): restaurant service, food taste, health and hygiene, variety and eating manners, and unique cultural experience
Yasami et al. (2020)	Four dimensions (22 items): restaurant service, food safety and health, food taste and popularity, variety, and table manners
Yılmaz et al. (2020)	Four dimensions (22 items): gastronomic identity, diversity of the destination, gastronomic attractions, and qualified workforce and stakeholders
Mohamed et al. (2021)	Nine dimensions (32 items): health and nutrition, food culture and heritage, distinctiveness, food quality and variety, flavor profile, cooking methods and ingredients, price and informality, dining places, and table manners

Table 11.6 Items of food affective image

Author (year)	*Items of food affective image*
Peštek and Cinjarevic (2014)	Boring – exciting, sleepy – arousing
Seo and Yun (2015)	Exciting, pleasant, fulfilled, enjoyable, and contented
Lai et al. (2019)	Excited, pleasant, satisfied, aroused, enjoyable, fun, good, and relax

Table 11.7 Items and dimensions of food cognitive image

Author (year)	*Items of food conative image*
Chi et al. (2013), Karim et al. (2009), and Peštek and Cinjarevic (2014)	Tourist satisfaction with the food experience
Lertputtarak (2012)	Revisit intention
Karim and Chi (2010) and Kim et al. (2012)	Visit intention
Sanchez-Cañizares and Castillo-Canalejo (2015)	Tourist satisfaction with the food experience and the willingness to pay more to try traditional food
Seo and Yun (2015)	Destination food preference and intention to eat destination foods
Lai et al. (2019)	Recommend intention

with the food experience (Chi et al., 2013), destination food preference and intention to eat destination foods (Seo & Yun, 2015), the willingness to pay more to try traditional food (Sanchez-Cañizares & Castillo-Canalejo, 2015), and recommend intention (Lai et al., 2019) (Table 11.7).

Relationships between destination food image and other concepts

Presently, studies on the relationships between destination food image and other variables in the tourism field are quite rare. The main focus is to explore the consequences of destination food image, that is, the influence on tourists' behavioral intention. Although many scholars disagree on the measurement criteria for destination food image, they all recognize the fact that destination food image has a certain influence on the tourists' behavioral intention, such as food preference, food expectation, food satisfaction, food destination loyalty, recommendation intention, and revisit intention (Seo et al., 2017; Toudert & Bringas-Rabago, 2019; Yasami et al., 2020). When Lertputtarak (2012) explored the destination food image of Thailand, he found that it was positively correlated with the revisit intention.

Chi et al. (2013) noticed that the destination food image not only directly affects the tourists' behavior intentions but also indirectly influences the tourists' behavior intentions through food satisfaction and food quality. Peštek and Cinjarevic (2014) analyzed the tourists' perception of Bosnian culture from both cognitive and affective perspectives and discovered that the more positive the tourists' perception of destination food image was, the more likely they were to have higher satisfaction with food experience and the more inclined they were to visit Bosnia again and recommend Bosnia to others. Toudert and Bringas-Rabago (2019) pointed out that tourists and excursionist have several differences in their perception of food cognitive image, but both have significant impacts on tourists' satisfaction and future intention.

Rare antecedents have been studied, including only nationality (Promsivapallop & Kannaovakun, 2019), tourists' attitudes toward local food (Choe & Kim, 2018), product quality, food quality, food value, and food diversity (Toudert & Bringas-Rabago, 2019). What's more, it is worth paying attention to that; some scholars have also noticed that there may be relationships between tourists' food-related personality traits and destination food image. For instance, Lai et al. (2019) testified that food neophobia significantly moderated the formation of destination food image.

Affective image is based on the cognitive image, and it has become a consensus that destination cognitive image significantly affects destination affective image (Beerli & Martin, 2004; Gartner, 1994; Lin et al., 2007). Accordingly, besides the relationships with external variables, the internal mechanism of destination food image also deserves attention. Lai et al. (2019) explored the relationships between the various dimensions of the Australian food image based on the cognitive-affective-conative model and found that both food cognitive image and affective image positively influence food conative image, and the positive effect of food cognitive image on food conative image is mediated by food affective image.

Conclusion

Food tourism is a continuously developing field. The purpose of this chapter is to review the literature on destination food image, analyze its research status, and explore new and valuable future research perspectives. With the continuous improvement of China's comprehensive national strength and national income, outbound tourism has gradually become an important consumption area for Chinese residents. At the same time, China's outbound tourism has become an economic and cultural phenomenon attracting worldwide attention and is constantly affecting the operation pattern of China's tourism economy and the trend of tourism market in destination countries or regions. According to data issued by the China Tourism Academy, China's outbound tourism market reached 155 million people, and outbound tourists spent more than US$133.8 billion abroad in 2019.[5]

Thus, Chinese tourists have become a market group that governments and tourism departments of all countries must pay attention to. Understanding their behavior is critical to the development of the destinations. Furthermore, market segments are becoming more and more popular in outbound tourism. The study focusing on a market segment is of great contribution to the in-depth understanding of the complex transnational consumption behavior of outbound tourists. As a segmented market, food tourism is a reflection of local culture and locality and can meet the differentiated and unique needs of specific tourist groups (George, 2019), so it is widely favored by Chinese outbound tourists.

Scholars seldom study the destination food image perceived by Chinese outbound tourists. Promsivapallop and Kannaovakun (2019) conducted a survey of Chinese and Australian tourists to Phuket and found that destination food image is not only influenced by nationality but also significantly affects food preference and consumption behavior. Lai et al. (2019) confirmed that the Australian food image perceived by Chinese tourists and food neophobia have positive impacts on their visit intention. Yasami et al. (2020) believe that the Phuket food image perceived by Chinese tourists is mainly divided into four dimensions, namely restaurant service, food safety and health, food taste and popularity, and variety and table, and discovered the direct effect of destination food image on destination loyalty behavior and the indirect influence produced by food satisfaction. Therefore, it is essential to further expand the research scope of destination food image perceived by Chinese outbound tourists and make international comparison in the future.

As mentioned above, the current studies on the relationships between destination food image and other variables are few and single, which can be further expanded. For example, it is possible to test the relationships between food tourism motivation and food destination image. Various research on the measurement of food tourism motivation have been proposed these years, but the relationships between it and other variables, especially the destination food image, have not been further examined. Moreover, the possible impact of other food-related personality traits on destination food image can also be explored such as need for uniqueness.

Need for uniqueness, as a psychological concept, refers to the individual uniqueness that people constantly seek to be recognized and appreciated by others (Snyder, 1992). In fact, everyone desires uniqueness more or less, which makes the need for uniqueness a very common personality trait (Burns & Brady, 1992). Especially with the further development of the experience-based economy, tourists' needs have also begun to change, and they are more pursuing uniqueness and individuality. According to the 2018 Consumer Research Report released by Hilton and Nielsen, among Chinese respondents, up to 77% of the group prefer unique consumer experiences and look forward to changes and surprises.[6] And the 2018 China Outbound Travel Shopping White Paper jointly issued by the Moody's David Report, and IClick also found that the post-90s generation pays more attention to unique and exclusive.[7]

The 2020 China Consumer Survey Report presented by McKinsey shows that products and services with more unique and memorable experiences can make increasingly mature Chinese consumers feel happy and surprised.[8] Consequently, unique local characteristics are more attractive to contemporary tourists, especially young groups, and destinations with unique food are more widely favored by tourists. Besides, food tourism itself is a niche tourism, whose tourist groups usually oppose homogenized and undifferentiated mass tourism products and pursue uniqueness (Sidali et al., 2015). Accordingly, it is necessary to explore whether there are group differences in destination food image perceived by tourists with different degrees of need for uniqueness.

As a personality trait, need for uniqueness gradually extends from the field of psychology to the field of consumers. The research is quite extensive not only in many application scenarios but also in detailed research on its antecedents and consequences.

Tian et al. (2001, p. 52) defined the need for uniqueness of consumers as "the trait of pursuing differentness relative to others through the acquisition, utilization, and disposition of consumer goods for the purpose of developing and enhancing one's self-image and social image" and created the corresponding scale of consumers' need for uniqueness (CNFU) based on previous studies. Later, in order to better explore the application of need for uniqueness in foods and beverages, Cardello et al. (2019) developed a special scale of need for uniqueness in foods and beverages (FBNFU) based on CNFU (Table 11.8).

More importantly, in spite of numerous research of need for uniqueness in the consumer field, there is no research on the impact of need for uniqueness on tourists' behavioral intention in the tourism academia. Only one article considered the need for uniqueness but only explored the catering consumption behavior during the process of travel, without discussing the tourist' behavioral intention. Hyun and Park (2016) cited the CNFU scale to explore that when tourists are dining out, their need for uniqueness is affected by perceived firm innovativeness, perceived firm uniqueness and brand prestige, and the need for uniqueness has significant impacts on utilitarian value, hedonic value, and behavioral intentions. And the FBUFU scale has not yet been applied in the tourism field. Considering that this scale is highly compatible with food tourism, the need for uniqueness can be introduced into the tourism field in the future, and FBNFU scale can be combined to explore destination food image perceived by food tourists and their behavioral intention.

Table 11.8 Items and dimensions of need for uniqueness

Author (year)	*Items and dimensions of need for uniqueness*
Snyder and Fromkin (1977)	32 items
Lynn and Harris (1997)	Four items
Tian et al. (2001)	Three dimensions (31 items): creative choice counter conformity, avoidance of similarity, and unpopular choice counter conformity
Ruvio et al. (2008)	Three dimensions (12 items): creative choice counter conformity, avoidance of similarity, and unpopular choice counter conformity
Rajamma et al. (2010)	Three dimensions (20 items): creative choice counter conformity, avoidance of similarity, and unpopular choice counter conformity
Cardello et al. (2019)	31 items

Notes

1 Travel weekly, 2018.
2 Booking.com: Eight Travel Trends for 2020 Report, 2019.
3 World Food Travel Association: Food travel monitor, 2016.
4 China Tourism Academy & Ctrip: 2017 Big Data Report on China's Outbound Tourism, 2018.
5 China Tourism Academy, 2019.
6 Hilton & Nielsen: The New Consumer Research Report, 2018.
7 IClick, Moody's David Report: White Paper on Chinese Outbound Travel Shopping, 2018.
8 McKinsey: China Consumer Survey Report 2020, 2020.

References

Agyeiwaah, E., Otoo, F., Suntikul, W., & Huang, W. (2019). Understanding culinary tourist motivation, experience, satisfaction, and loyalty using a structural approach. *Journal of Travel & Tourism Marketing*, *36*(3), 295–313. doi:10.1080/10548408.2018.1541775

Antón, C., Camarero, C., Laguna, M., & Buhalis, D. (2019). Impacts of authenticity, degree of adaptation and cultural contrast on travellers' memorable gastronomy experiences. *Journal of Hospitality Marketing & Management*, *28*(7), 743–764. doi:10.1080/19368623.2019.1564106

Baloglu, S., & McCleary, K. W. (1999). A model of destination image formation. *Annals of Tourism Research*, *26*(4), 868–897. doi:10.1016/s0160–7383(99)00030-4

Beerli, A., & Martín, J. D. (2004). Tourists' characteristics and the perceived image of tourist destinations: A quantitative analysis—a case study of Lanzarote, Spain. *Tourism Management*, *25*(5), 623–636. doi:10.1016/j.tourman.2003.06.004

Björk, P., & Kauppinen-Räisänen, H. (2016). Local food: A source for destination attraction. *International Journal of Contemporary Hospitality Management*, *28*(1), 177–194. doi:10.1108/ijchm-05–2014–0214

Boniface, P. (2003). *Tasting tourism: Travelling for food and drink*. London: Routledge. doi:10.4324/9781315241777

Boyne, S., Hall, D., & Williams, F. (2003). Policy, support and promotion for food-related tourism initiatives. *Journal of Travel & Tourism Marketing*, *14*(3–4), 131–154. doi:10.1300/j073v14n03_08

Burns, D., & Brady, J. (1992). A cross-cultural comparison of the need for uniqueness in Malaysia and the United States. *The Journal of Social Psychology*, *132*(4), 487–495. doi:10.1080/00224545.1992.9924728

Cardello, A., Chheang, S., Hedderley, D., Guo, L., Hunter, D., & Jaeger, S. (2019). Toward a new scale to measure consumers' "need for uniqueness" in foods and beverages: The 31-item FBNFU scale. *Food Quality and Preference*, *72*, 159–171. doi:10.1016/j.foodqual.2018.10.008

Çela, A., Knowles-Lankford, J., & Lankford, S. (2007). Local food festivals in Northeast Iowa communities: A visitor and economic impact study. *Managing Leisure*, *12*(2–3), 171–186. doi:10.1080/13606710701339470

Chaigasem, T., & Tunming, P. (2019). A local cuisine tourism approach to authenticity and a sense of place for Postmodern gastronomy in I-SAN Thailand. *African Journal of Hospitality, Tourism and Leisure*, 8(4), 1–14. https://www.ajhtl.com/uploads/7/1/6/3/7163688/article_29_vol_8__4__2019_thailand.pdf

Chen, C.-F., & Phou, S. (2013). A closer look at destination: Image, personality, relationship and loyalty. *Tourism Management, 36*, 269–278. doi:10.1016/j.tourman.2012.11.015

Chi, C. G.-Q., Chua, B. L., Othman, M., & Karim, S. A. (2013). Investigating the structural relationships between food image, food satisfaction, culinary quality, and behavioral intentions: The case of Malaysia. *International Journal of Hospitality & Tourism Administration, 14*(2), 99–120. doi:10.1080/15256480.2013.782215

Cho, S. (2019). Effects of food gourmet tourism motivation and satisfaction on recommendation of the degree of involvement in gastronomy. *Journal of Cultural Product & Design, 56*, 39–48.

Choe, J. Y. (Jacey), & Kim, S. (Sam). (2018). Effects of tourists' local food consumption value on attitude, food destination image, and behavioral intention. *International Journal of Hospitality Management, 71*, 1–10. doi:10.1016/j.ijhm.2017.11.007

Crompton, J. L. (1979). An assessment of the image of Mexico as a vacation destination and the influence of geographical location upon that image. *Journal of Travel Research, 17*(4), 18–23. doi:10.1177/004728757901700404

Dann, G. M. S. (1977). Anomie, ego-enhancement and tourism. *Annals of Tourism Research, 4*(4), 184–194. doi:10.1016/0160–7383(77)90037-8

Di-Clemente, E., Hernández-Mogollón, J., & Campón-Cerro, A. (2020). Tourists' involvement and memorable food-based experiences as new determinants of behavioural intentions towards typical products. *Current Issues in Tourism, 23*(18), 2319–2332. doi:10.1080/13683500.2019.1631265

Ellis, A., Park, E., Kim, S., & Yeoman, I. (2018). What is food tourism?. *Tourism Management, 68*, 250–263. doi:10.1016/j.tourman.2018.03.025

Fields, K. (2002). Demand for the gastronomy tourism product: Motivational factors. In A. M. Hjalager & G. Richards (Eds.), *Tourism and gastronomy* (pp. 36–50). London: Routledge.

Gartner, W. C. (1994). Image formation process. *Journal of Travel & Tourism Marketing, 2*(2–3), 191–216. doi:10.1300/j073v02n02_12

George, R. (2019) *Marketing tourism in South Africa*. Cape Town: Oxford University Press.

Hall, C., & Sharples, L. (2003). The consumption of experiences or the experiences of consumption? An introduction to the tourism of taste. In C. M. Hall, E. Sharples, R. Mitchell, N. Macionis, & B. Cambourne (Eds.), *Food tourism around the world: Development, management and markets* (pp. 1–24). Oxford: Butterworth-Heinemann.

Hillel, D., Belhassen, Y., & Shani, A. (2013). What makes a gastronomic destination attractive? Evidence from the Israeli Negev. *Tourism Management, 36*, 200–209. doi:10.1016/j.tourman.2012.12.006

Hjalager, A. (2003). What do tourists eat and why? Towards a sociology of gastronomy and tourism. *CAUTHE 2003: Riding the Wave of Tourism and Hospitality Research*, pp. 546–561. https://search.informit.org/doi/abs/10.3316/informit.841777037489926

Horng, J.-S., Liu, C.-H., Chou, H.-Y., & Tsai, C.-Y. (2012). Understanding the impact of culinary brand equity and destination familiarity on travel intentions. *Tourism Management, 33*(4), 815–824. doi:10.1016/j.tourman.2011.09.004

Horng, J.-S., Su, C.-S., & So, S.-I. A. (2013). Segmenting food festival visitors: Applying the theory of planned behavior and lifestyle. *Journal of Convention & Event Tourism, 14*(3), 193–216. doi:10.1080/15470148.2013.814038

Horng, J.-S., & Tsai, C.-T. (Simon) (2010). Government websites for promoting East Asian culinary tourism: A cross-national analysis. *Tourism Management, 31*(1), 74–85. doi:10.1016/j.tourman.2009.01.009

Hunter, W. C. (2012). Projected destination image: A visual analysis of Seoul. *Tourism Geographies, 14*(3), 419–443. doi:10.1080/14616688.2011.613407

Hyun, S. S., & Park, S. H. (2016). The antecedents and consequences of travelers' need for uniqueness: An empirical study of restaurant experiences. *Asia Pacific Journal of Tourism Research, 21*(6), 596–623. doi:10.1080/10941665.2015.1062404

Ignatov, E., & Smith, S. (2006). Segmenting Canadian culinary tourists. *Current Issues in Tourism, 9*(3), 235–255. doi:10.2167/cit/229.0

Karim, S., & Chi, C. G.-Q. (2010). Culinary tourism as a destination attraction: An empirical examination of destinations' food image. *Journal of Hospitality Marketing & Management, 19*(6), 531–555. doi:10.1080/19368623.2010.493064

Karim, S., Chua, B., & Salleh, H. (2009). Malaysia as a culinary tourism destination: International tourists' perspective. *Journal of Tourism, Hospitality and Culinary Arts, 1*(3), 63–78. https://fhtm.uitm.edu.my/images/jthca/Vol1Issue3/Chap-4.pdf

Kim, S., Kim, M., Agrusa, J., & Lee, A. (2012). Does a food-themed TV drama affect perceptions of national image and intention to visit a country? An empirical study of Korea TV drama. *Journal of Travel & Tourism Marketing, 29*(4), 313–326. doi:10.1080/10548408.2012.674869

Kim, Y. G., & Eves, A. (2012). Construction and validation of a scale to measure tourist motivation to consume local food. *Tourism Management, 33*(6), 1458–1467. doi:10.1016/j.tourman.2012.01.015

Kim, Y. G., Eves, A., & Scarles, C. (2009). Building a model of local food consumption on trips and holidays: A grounded theory approach. International *Journal of Hospitality Management, 28*(3), 423–431. doi:10.1016/j.ijhm.2008.11.005

Kivela, J., & Crotts, J. (2005). Gastronomy tourism: A meaningful travel market segment. *Journal of Culinary Science & Technology, 4*(2–3), 39–55. doi:10.1300/j385v04n02_03

Kivela, J., & Crotts, J. C. (2006). Tourism and gastronomy: Gastronomy's influence on how tourists experience a destination. *Journal of Hospitality & Tourism Research, 30*(3), 354–377. doi:10.1177/1096348006286797

Lai, M. Y., Khoo-Lattimore, C., & Wang, Y. (2017). Food and cuisine image in destination branding: Toward a conceptual model. *Tourism and Hospitality Research, 19*(2), 238–251. doi:10.1177/1467358417740763

Lai, M. Y., Wang, Y., & Khoo-Lattimore, C. (2019). Do food image and food neophobia affect tourist intention to visit a destination? The case of Australia. *Journal of Travel Research*, 1–22. doi:10.1177/0047287519867144

Lertputtarak, S. (2012). The relationship between destination image, food image, and revisiting Pattaya, Thailand. *International Journal of Business and Management, 7*(5), 111–222. doi:10.5539/ijbm.v7n5p111

Levitt, J. A., Meng, F., Zhang, P., & DiPietro, R. B. (2019). Examining factors influencing food tourist intentions to consume local cuisine. *Tourism and Hospitality Research, 19*(3), 337–350. doi:10.1177/1467358417742687

Lin, C.-H., Morais, D. B., Kerstetter, D. L., & Hou, J.-S. (2007). Examining the role of cognitive and affective image in predicting choice across natural, developed, and theme-park destinations. *Journal of Travel Research, 46*(2), 183–194. doi:10.1177/0047287507304049

Lin, Y.-C., Pearson, T. E., & Cai, L. A. (2011). Food as a form of destination identity: A tourism destination brand perspective. *Tourism and Hospitality Research*, *11*(1), 30–48. doi:10.1057/thr.2010.22

Long, L. (2004). *Culinary tourism*. Lexington: The University Press of Kentucky.

Lynn, M., & Harris, J. (1997). The desire for unique consumer products: A new individual differences scale. *Psychology and Marketing*, *14*(6), 601–616. doi:10.1002/(sici)1520–6793(199709)14:6<601::aid-mar5>3.0.co;2-b

Mak, A. H. N., Lumbers, M., Eves, A., & Chang, R. C. Y. (2012). Factors influencing tourist food consumption. *International Journal of Hospitality Management*, *31*(3), 928–936. doi:10.1016/j.ijhm.2011.10.012

McKercher, B., Okumus, F., & Okumus, B. (2008). Food tourism as a viable market segment: It's all how you cook the numbers! *Journal of Travel & Tourism Marketing*, *25*(2), 137–148. doi:10.1080/10548400802402404

Meler, M., & Cerovic′, Z. (2003). Food marketing in the function of tourist product development. *British Food Journal*, *105*(3), 175–192. doi:10.1108/00070700310477121

Mohamed, M. E., Lehto, X., Hewedi, M., & Behnke, C. A. (2021). Naïve destination food images: Exploring the food images of non-visitors. *Journal of Hospitality and Tourism Management*, *47*, 93–103. doi:10.1016/j.jhtm.2021.03.005

Nicoletti, S., Medina-Viruel, M. J., Di-Clemente, E., & Fruet-Cardozo, J. V. (2019). Motivations of the culinary tourist in the city of Trapani, Italy. *Sustainability*, *11*(9), 2686. doi:10.3390/su11092686

Peštek, A., & Činjarević, M. (2014). Tourist perceived image of local cuisine: The case of Bosnian food culture. *British Food Journal*, *116*(11), 1821–1838. doi:10.1108/bfj-01–2014–0046

Pike, S., & Ryan, C. (2004). Destination positioning analysis through a comparison of cognitive, affective, and conative perceptions. *Journal of Travel Research*, *42*(4), 333–342. doi:10.1177/0047287504263029

Promsivapallop, P., & Kannaovakun, P. (2019). Destination food image dimensions and their effects on food preference and consumption. *Journal of Destination Marketing & Management*, *11*, 89–100. doi:10.1016/j.jdmm.2018.12.003

Quan, S., & Wang, N. (2004). Towards a structural model of the tourist experience: An illustration from food experiences in tourism. *Tourism Management*, *25*(3), 297–305. doi:10.1016/s0261–5177(03)00130-4

Rajamma, R. K., Pelton, L. E., Hsu, M. K., & Knight, D. K. (2010). The impact of consumers' need for uniqueness and nationality on generation Y's retail patronage behaviors: Investigating American and Taiwanese consumers. *Journal of Global Marketing*, *23*(5), 387–410. doi:10.1080/08911762.2010.521112

Robinson, R. N. S., Getz, D., & Dolnicar, S. (2018). Food tourism subsegments: A data-driven analysis. *International Journal of Tourism Research*, *20*(3), 367–377. doi:10.1002/jtr.2188

Ruvio, A., Shoham, A., & Makovec Brenčič, M. (2008). Consumers' need for uniqueness: Short-form scale development and cross-cultural validation. *International Marketing Review*, *25*(1), 33–53. doi:10.1108/02651330810851872

Sanchez-Cañizares, S., & Castillo-Canalejo, A. M. (2015). A comparative study of tourist attitudes towards culinary tourism in Spain and Slovenia. *British Food Journal*, *117*(9), 2387–2411. doi:10.1108/bfj-01–2015–0008

Santich, B. (2004). The study of gastronomy and its relevance to hospitality education and training. *International Journal of Hospitality Management*, *23*(1), 15–24. doi:10.1016/s0278–4319(03)00069-0

Seo, S., & Yun, N. (2015). Multi-dimensional scale to measure destination food image: Case of Korean food. *British Food Journal*, *117*(12), 2914–2929. doi:10.1108/bfj-03–2015–0114

Seo, S., Yun, N., & Kim, O. Y. (2017). Destination food image and intention to eat destination foods: A view from Korea. *Current Issues in Tourism*, *20*(2), 135–156. doi:10.1080/13683500.2014.934210

Sidali, K. L., Kastenholz, E., & Bianchi, R. (2015). Food tourism, niche markets and products in rural tourism: Combining the intimacy model and the experience economy as a rural development strategy. *Journal of Sustainable Tourism*, *23*(8–9), 1179–1197. doi:10.1080/09669582.2013.836210

Sims, R. (2009). Food, place and authenticity: Local food and the sustainable tourism experience. *Journal of Sustainable Tourism*, *17*(3), 321–336. doi:10.1080/09669580802359293

Smith, S., Costello, C., & Muenchen, R. A. (2010). Influence of push and pull motivations on satisfaction and behavioral intentions within a culinary tourism event. *Journal of Quality Assurance in Hospitality & Tourism*, *11*(1), 17–35. doi:10.1080/15280080903520584

Snyder, C. R. (1992). Product scarcity by need for uniqueness interaction: A consumer Catch-22 carousel?. *Basic and Applied Social Psychology*, *13*(1), 9–24. doi:10.1207/s15324834basp1301_3

Snyder, C. R., & Fromkin, H. L. (1977). Abnormality as a positive characteristic: The development and validation of a scale measuring need for uniqueness. *Journal of Abnormal Psychology*, 86(5), 518–527. doi: 10.1037/0021–843X.86.5.518

Su, D. N., Johnson, L. W., & O'Mahony, B. (2020). Analysis of push and pull factors in food travel motivation. *Current Issues in Tourism*, *23*(5), 572–586. doi:10.1080/13683500.2018.1553152

Tasci, A. D. A., Gartner, W. C., & Cavusgil, S. T. (2007). Conceptualization and operationalization of destination image. *Journal of Hospitality & Tourism Research*, *31*(2), 194–223. doi:10.1177/1096348006297290

Tian, K. T., Bearden, W. O., & Hunter, G. L. (2001). Consumers' need for uniqueness: Scale development and validation. *Journal of Consumer Research*, *28*(1), 50–66. doi:10.1086/321947

Toudert, D., & Bringas-Rábago, N. L. (2019). Destination food image, satisfaction and outcomes in a border context: Tourists vs excursionists. *British Food Journal*, *121*(5), 1101–1115. doi:10.1108/bfj-03–2019–0148

Tsai, C.-T. (Simon), & Wang, Y.-C. (2017). Experiential value in branding food tourism. *Journal of Destination Marketing & Management*, *6*(1), 56–65. doi:10.1016/j.jdmm.2016.02.003

Yasami, M., Promsivapallop, P., & Kannaovakun, P. (2020). Food image and loyalty intentions: Chinese tourists' destination food satisfaction. *Journal of China Tourism Research*, 1–21. doi:10.1080/19388160.2020.1784814

Yılmaz, G., Kılıçarslan, D., & Caber, M. (2020). How does a destination's food image serve the common targets of the UNESCO creative cities network? *International Journal of Tourism Cities*, *6*(4), 785–812. doi:10.1108/ijtc-07–2019–0115

12 Local perceptions and behavior of Chinese tourists at the tulip fields in the Netherlands

Xiaodan Zhao and Nitasha Sharma

Introduction

Several cities in Europe such as Venice, Amsterdam, and Barcelona have faced the negative impacts of overtourism for years. A discussion of sustainable tourism and responsible tourist behavior has become common in social media and academic discourse since tourism could cause negative environmental, cultural, and social impacts on the lives of residents (Kang & Moscardo, 2006). Thus, several popular tourist destinations are transforming from destination marketing to destination management, employing strategies for managing tourist behavior. Cheese, clogs, windmills, and tulip flowers are the most recognizable icons of tourism in the Netherlands. Tourists can enjoy most treats and sights around the year but can only enjoy the tulip flowers for a few weeks during the Spring season. However, overtourism in the Netherlands has been a persistent problem, especially in Amsterdam and nearby areas with hordes of tourists being particularly disruptive at the tulip fields.

Tourist behavior and performance at the tulip fields became a matter of concern over the past few years, leading to potentially adverse environmental and social impacts. The farmers face financial losses every year because tourists trample over the tulips for taking photos. The tourism authorities often blame social media for fueling the trespassing problem. Also, several international media outlets reported transgressive behavior by international tourists at the tulip fields which indirectly contributed to a stereotyped image of misbehaving Chinese tourists. As reported by media outlets such as *The Guardian*, *NL Times*, and *The Dutch Review*, several signs were put up in the tulip fields, reading, 'Enjoy the flowers, respect our pride', in English and Chinese to remind the tourists not to lie down in the flower fields or stomp on it for that perfect selfie. The media reports slowly started associating Chinese tourists with negative terms like 'misbehavior', 'rude', or 'invasion'.

A report in 2019 by *The Guardian* highlighted how Dutch farmers used signs and banners at the tulip fields to control tourist behavior and spread awareness. The fact that these signs were written in English and Chinese/Mandarin raises essential questions about Chinese tourists' behavior at the

DOI: 10.4324/9781003121329-14

tulip fields and local perceptions. Therefore, the objective of this study is to explore the motivations, perceptions, and behavior of Chinese tourists at the tulip fields in the Netherlands, along with local perspectives on tourism and tourist behavior. A multi-method research approach comprising qualitative methods such as semi-structured interviews, questionnaires, and visual discourse analysis of tourist photographs has been used to provide an insight into the marketing of tulip fields, the psychology behind Chinese tourist transgressions, and the organization of tulip-related activities for the tourists. It attempts to understand the role of tulip fields in the broader destination image of the Netherlands as constructed by Chinese tourists and how Chinese tourists perceive transgressive behavior at the tulip fields.

The study is significant as it contributes to a better understanding of tourist behavior management at the tulip fields. It aims at reducing the reproduction and circulation of racial tourist stereotyping of Chinese visitors and reducing cultural friction between Chinese tourists and Dutch locals. As a theoretical framework, the notion of sense of place, host-guest relations in the context of overtourism, and moral behavior in tourism studies have been used. Despite discussions regarding organizing tourist behavior from a management perspective, few studies have sought to explain the tourists' behavior through psychological theories. The process of conducting moral action is complicated for the tourists in cultural settings different from their own as it relies on several factors. Hence, exploring the Chinese tourists' motivations, expectations, and behavior, based on the above theoretical ideas, can provide an insight into what prompts tourist transgressions at the tulip fields as well as their experiences.

Tulip tourism

A popular activity for inbound tourists every year is to visit the famous Keukenhof tulip garden in the Netherlands. Since it is the most popular attraction on Instagram, the crowd of tourists also spreads out to Bollenstreek and other nearby tulip fields around Keukenhof. The tulip gardens are distributed from the northern part of the Netherlands, such as Groningen, to the southern part, like Den Haag. The over 100 km long 'Flower bulb route', through 2,500 acres of tulip fields, was nominated as one of the most beautiful flower routes by National Geographic. Every year, approximately more than 1.7 billion tulips are exported for business purposes. Several events and festivals such as the National Tulip Flower Day are also organized to attract local and international tourists. The Tulip festival that lasted three weeks in Noordoostpolder, Flevoland province, received 150,000 tourists in 2019.

Although Chinese tourists are not the majority of the visitors compared to the European market, the Netherlands has been a popular destination for Chinese tourists who mainly visit during springtime for enjoying the tulip flowering season. The guided tours remain popular with group sizes getting smaller as those traveling with their family and the young generation prefer to choose the fully independent tour and semi-independent travel (only including flight tickets and hotels). With the rapid growth of the Chinese outbound tourists' market in the global world, more and more Chinese tourist misbehavior has been reported in social media and newspapers. This is a point of concern as tourists' misbehavior influences other tourists' experience and providers' service, even the reputation of the country of their origin.

Methodology

The areas such as 'Bollenstreek' (meaning 'flower bulb region' in Dutch) are popular on social media for their rural landscape views with colorful fields. Additional tulip fields beyond Keukenhof are accessible by bikes or cars and less commercialized such as Noordoostpolder in Flevoland province. These beautiful locations are suitable for taking photos and are frequently recommended on the internet. The data for this study were collected from March to June 2020, and local interviewees from Flevoland and Bollenstreek were selected by contacting them online and over the phone and respondents were contacted through acquaintances, social media tags, and check-ins of tourists who have visited those areas. The study adopted a multi-method approach for the study including the collection of qualitative data, analysis of secondary data, literature review, narratives collected using semi-structured interviews, online surveys, and visual discourse analysis.

The secondary sources of data and literature included background information concerning the tulip fields, domestic and international tourist numbers, the history behind the cultivation of tulips in this geographical area, and reports of Chinese tourists' misbehavior. Tourists who had visited the tulip fields within the last two years or plan to visit the fields were selectively chosen from personal contacts and social media. The respondents were recruited through the snowball technique, where the primary respondents passed on the contact details of their friends and acquaintances who had visited the tulip fields earlier. Before the interviews, informed consent was obtained from the respondents, and the goals of the study were explained to them. The data was thematically processed on ATLAS. ti 8. The social media apps such as Dian Ping (Chinese Yelp), TripAdvisor, and Instagram were analyzed for geotagged photos that tourists posted on tulip field sightseeing trips to note how frequently certain types of images appeared. The visual discourse based on tourist photos is part of the cultural narrative that tourism marketing officials have promoted globally.

The questionnaires consisted of open-ended questions attempting to understand how Chinese tourists perceive the Netherlands, source of travel information, their expectations, activity plans, motivation toward visiting the tulip fields, experiences, their level of awareness, and the possibility of engaging in transgressive behavior at the tulip fields. They were provided in both Chinese and English. The collected data were processed anonymously, and 100 valid responses from tourists were collected.

Semi-structured interviews were conducted with local tulip field growers and stakeholders. The goals of the semi-structured interviews were to assess strategies of management related to the tulip fields and local perceptions about the tourism industry and tourists' behavior. One of the respondents was the regional manager and the chairman of Royal General Association for Flower Bulb Culture (KAVB) in the Flevoland provincial branch. KAVB has been an independent agricultural organization in the Netherlands since 1860; they hold the Tulip Festival in Noordoostpolder every year, which is very popular on social media. Another prominent respondent was the coordinator of the Flower Science organization (NGO) and the initiator of the tourist campaign in Bollenstreek. Another interview was conducted with an officer from the tulip marketing campaign at the Netherlands Board of Tourism & Conventions (NBTC), a marketing organization that promotes tulip field sightseeing information globally.

Tourist motivations and perceptions

The tourist responses toward the tulip fields ranged from labeling the tulip fields as 'beautiful', 'exotic experience', 'Dutch tulips are famous' to 'relaxing experience', 'photogenic', 'romantic' (indicating wedding and engagement photo-shoots, walking through the fields with loved ones, etc.), and 'representing Dutch culture'. Based on the results obtained from the survey combined with the visual discourse analysis, the tourist responses can be categorized under four themes: the romantic gaze, pursuit of authenticity, psychological well-being, and playfulness.

The romantic tourist gaze

According to Urry (1990), visual consumption and the gaze are the most important tourist activities especially when it comes to nature or sightseeing. According to Urry and Larsen (2011):

> Places are chosen to be gazed upon because there is anticipation, especially through daydreaming and fantasy, of intense pleasures, either on a different scale or involving different senses from those customarily encountered. Such anticipation is constructed and sustained through a variety of non-tourist technologies, such as films, TV, literature, magazines, CDs, DVDs, and videos, constructing and reinforcing the gaze. (p. 4)

As stated by Li et al. (2019), many Chinese tourists favored traveling and engaging with the views which they have "encountered in a transcendental and interactive manner" (p. 518). This interaction and transcendence can be observed from the responses of the Chinese tourists where 'Take pictures', 'Romantic activities', and 'Tulips are beautiful', appeared as high-frequency keywords, thereby indicating an aesthetic and romantic perspective toward the tulip fields.

Chinese tourists are influenced by the promotional pictures in various information resources, reinforcing the rural image of the tulip field. Their desire of taking pictures can be understood as the search for that beauty as promoted by the marketing campaigns and advertisements in their home country. The aesthetic desire to capture the beauty of the tulip fields is partly rooted in the desire of Chinese tourists to gaze upon natural beauty and open green spaces, stimulated by the combination of environmental problems, excessive commercialization of natural and cultural heritage, and contemporary life in China characterized by high pressure, materialism, and fierce competition (Li et al., 2019). 'Romantic activities' also prove that the tulip field is not only a sightseeing spot; it is a symbol that stands for passionate feelings in personal relationships, which means that the tourists only want to share it with someone intimate.

Indeed, "part of what is involved in tourism is the purchase of a particular themed experience, and this depends upon a specifiable composition of the others with whom that experience is being shared" (Urry & Larsen, 2011, p. 70). The tulip fields reinforce that romantic theme among Chinese tourists who accordingly perceive the tulip field as a private space to escape from daily life pressures and release the emotions that they might not normally show otherwise.

Pursuit of authenticity

Besides complimenting on the beauty of the tulip fields, Chinese tourists commented that 'tulips are special, unique, and famous' and constitute an 'exotic experience'. These comments point toward Chinese tourists taking the tulip field as a unique and real representation of Dutch culture. The tulip is a distinctive element that makes their experience authentic. The tulips, in the marketing strategy, are constructed as an authentic cultural element by narratives, storytelling, and visual representation.

Constructive authenticity is a highly contextual term, a kind of authenticity that can be constructed by someone's dream or stereotypical image (Wang, 1999). Chinese tourists formulate a mental image of the tulip field before the trip, so visiting the tulip field is looking for the fulfillment of the expectation they had or checking whether it fits their constructed image. In other words, they are not considering the standards of authenticity in reality, but are, instead, capturing what they see, then transforming the meaning to fit their mental image. This idea is somewhat similar to the concept of imagined authenticity (Li et al., 2017), where tourists tend to stage/imagine their own authentic experience, even though it might be different from what the tourism industry in the host country wants them to experience. In this context, it can be indirectly inferred that the Chinese tourists prefer to borrow those elements of the Dutch landscape which are visually appealing, poetic, creative, or authentic while imagining and persuading themselves that they have had a glance at local culture and the authenticity of the toured place.

Psychological well-being

In the questionnaire, several Chinese tourists filled in 'relaxation' as the motivation for visiting the tulip fields or described their experience in terms of 'I feel relaxed' or being 'creative'. Tourists seeking feelings of relaxation can be understood by recourse to Maslow's hierarchy of needs which divides fundamental human needs into five stages: physiological, safety and security, love and belonging, esteem, and self-actualization (Maslow, 1943).

The notion of well-being has long been a trend in designing tourism products such as those of medical and nature-based tourism. Experiences that produce positive emotions for tourists, such as relaxation and happiness in the short term, also contribute to psychological well-being. Chinese tourists visualize the tulip fields with a romantic gaze and gain pleasure and relaxation from this activity. These emotional impacts form a positive aspect of their daily life in the long run. Further, being 'creative' means that tourists subconsciously link the feelings at the tulip field with self-actualization. Self-actualization is defined as "the full use and exploitation of one's talent, capacities, potentialities, etc." (Heylighen, 1992, p. 105). It shows that every individual has some potential they could discover or develop; these potentialities can be made actual, and the actualization can be conceived of as a continuous process of discovery. Self-actualization is also associated with creativity and acceptance. The tulip fields with different color combinations serve this purpose by triggering the feelings of creativity and inspiration among tourists.

Playfulness

Playfulness is an attitude that accompanies the individual from childhood to adulthood and can make adults distance themselves from their daily life (Kuo et al., 2016). It can be defined as "an internal predisposition characterized by creativity, curiosity, pleasure, sense of humor, and spontaneity" (Guitard et al., 2005, p. 9). In the context of tourism, playfulness can also be understood as the experience of emotions such as curiosity and the feeling of being inspired by sites (Kuo et al., 2016). The visual discourse analysis of photographs posted by Chinese tourists on social media revealed playfulness, joviality, a sense of creativity, and liveliness in their body language and postures.

The way tourists choose to take photos is influenced by their cultural aesthetics (Christmann, 2008). Accordingly, how Chinese tourists capture the tulip fields in their photos demonstrates how they appreciate the sites. The photos are flawless with various PS technics and color filters, being mainly about the colorful flower patterns (e.g., scenery photos), selfies with gestures (e.g., V-shape hand and heart-shaped hands with lovers), capturing a particular movement (e.g., jumping), and exotic props (e.g., wearing clogs in the field). Such content is interesting for viewers on social media, and the photos are usually shown in order, based on the number of positive reactions and browsing numbers. The display settings further stimulate tourists and travel bloggers to put more effort into taking good photos and edit their comments to earn more followers and viewers.

Role of tulips in destination image formation

Visual representations convey and shape the ideology and symbolic power involved in the construction of national identity; otherwise, these represented objects will remain invisible. The tourists were asked to describe three salient impressions of the Netherlands, and the words that appeared the most in order of highest frequency were 'windmill', 'tulip', and 'cheese', verifying the early marketing image promoted by the tourist officials. The perception of the tulip field as a rural, idyllic landscape is the result of the action and interaction of natural and/or human factors and represents a uniquely Dutch culture that plays an important role in destination image formation. The tulip field, as one of the national icons, and forming an aspect of the Dutch culture, can convey and transmit meaning and a particular message about the country. This narrative fits the promotional strategy on the national level as well as the local perspectives provided below:

> *We'll need to invest in a strong and sustainable promotion of our national pride, the flower fields, together with the sector and the flower industry.*
>
> [An official involved in the tulip campaign in NBTC]

> *In our area, we believe that flowers are associated with health and happiness. That's also is the feeling we will export to the world, that's our message, I guess.*
>
> [Local farmer]

According to the interviewees, the marketing strategy is to improve the residents' and tourists' understanding of national identity and pride and spread the positive meaning of tulip in the world, thereby culturally constructing the tulip as a symbol of national identity. The tulip as a cultural icon acts as a bridge linking the past and the present and the traditions and the people and enabling the Dutch population to be a distinctive group. The tulip field has a symbolic power creating a sense of emotional belongingness and national identity that naturally constitutes a sense of place and an affection for the past. Although the tulip is originally from Turkey, it has a close connection with the Netherlands since the end of the 16th century and reflects the historic trade-related prosperity. Over the years, tulips became integrated into the Dutch people's life as a social activity and an object of national pride. This partly explains why Chinese tourists consider the tulip field as a site that is obligatory for them to visit and experience it as bound up with a sense of place that symbolizes the Dutch people's life and their identity.

Transgressive behavior of Chinese tourists

Transgressive behavior could be interpreted differently depending on the context and different socio-cultural, political, and environmental settings. In the case of tulip fields, it can be defined as non-adherence to rules, regulations, and any form of written or non-written prohibitory sign or warning set up by the government, local authorities, farmers, and landowners that seek to restrict the following: (a) entry of outsiders into the fields, (b) plucking flowers, and (c) trampling over flowers. Further, transgressive behavior among tourists is best understood through a lens of ethics and moral decision-making.

Ethics in tourism is a contextually determined term with every cultural framework in society having a different understanding of ethics with no absolute right or wrong view. By contrast, moral absolutism means that a right or wrong view is dependent on universal principles, and moral decision-making is based on this perspective (Tolkach et al., 2017). Normative ethics is a branch of ethics that deals with such moral decision-making. In the context of tulip fields, an understanding of moral decision-making is essential to analyze why Chinese tourists may exhibit transgressive behavior.

To understand the moral decision-making of tourists at the tulip fields, we draw upon tourist psychology, in particular, as we regard moral disengagement as an essential element in tourists' individual- and collective-level decision-making process (Bandura, 1999). According to Bandura (2002, p. 102), "moral reasoning is linked to moral action through affective self-regulatory mechanisms by which moral agency is exercised". A disengagement of these self-regulatory mechanisms "mediates the relationship between the moral principles that individuals hold and their behavioral transgressions" (Sharma, 2020, p. 277). She adds that these mechanisms could be used in illustrating how the tourists adjust their assessment of their immoral action, for it to appear less immoral, thereby effectively shifting the moral boundaries.

The survey contained questions based on tourist behavior, about how tourists perceive themselves and other fellow tourists entering the tulip fields. The responses were determined using a Likert scale. When asked to what extent they are likely to enter the tulip field or pluck the flowers for their pleasure when no one else is around, the majority of the Chinese tourists replied 'No', only 2% replied 'Yes', and 6% demonstrated a possible response that they might pluck tulips when no one else is around. While 84% of tourists replied 'No', about 4% of tourists responded positively when asked to what extent they would follow others in stepping into the field, although not doing it first. This shows that they are likely to display transgressive behavior when they see their fellow tourists doing so.

The above results are in line with what Sharma (2020) mentioned about the moral disengagement mechanism 'diffusion of responsibility' which states that "transgressive acts are attributed to group decisions instead of personal agency. This way, personal agency is deliberately obscured and exercise of moral control is weakened exonerating the tourist from any form of personal responsibility or self-blame". According to Bandura (2002), individuals' standards of right or wrong are flexible when used as a guideline for conduct or in shaping moral agency. This explains why Chinese tourists tend to break rules especially if they are in large tour groups or they see other tourists doing so. Further, tourists have moral standards from their cultural framework, and they will judge their actions according to those standards and circumstances and then regulate their actions by following consequences.

Moral disengagement allows tourists an excuse to convince themselves and others that they are not guilty of their unethical actions. In Sharma (2020), this point is considered as similar to the process of deindividuation in social psychology, with application to situations when tourists act as a group or when other group members' behavior affects individual decision-making. The idea is that the sense of individuality and self-identity will, in a sense, be diminished. This is likely to give rise to the sense that responsibility will not attach itself to a specific person, and "any harm done by a group can always be attributed largely to the behavior of others" (Bandura et al., 1996, p. 290). Therefore, this explains that when an unethical action is operated by a collective decision, the Chinese tourists feel less guilty when they follow others into the tulip field. For example, when asked to what extent tourists would accept other people stepping into a tulip field for taking pictures to post online, around 26% of the tourists replied "Yes".

Further, when asked for the extent tourists would remind others not to step into the field when there is an entry-forbidden sign, 52% of tourists interestingly replied "Yes". This means that while tourists themselves may indulge in transgressive behavior, they demonstrate a likelihood to stop others from stepping into the field. When probed further, interviews revealed that Chinese tourists have a stereotyped notion that the Netherlands is a relatively free country due to the legalization of certain activities (such as smoking cannabis and prostitution) which are otherwise banned or illegal in their own or other countries, so they often face a struggle between self-sanction and social-sanction in this obscuring situation. This explains why some tourists answered "No".

However, the fact that 52% of tourists replied "Yes" fits with Bandura's (1991) argument that

> the exercise of moral agency has dual aspects—inhibitive and proactive. While the inhibitive form is manifested in the power to refrain from

> indulging in transgressive behavior, the proactive form of morality is expressed in the power to behave otherwise.
>
> (p. 194)

This was also pointed out by Sharma (2020) and might explain why certain Chinese tourists exercise their moral agency and decide to remind fellow tourists if the latter step into a tulip field with an entry-forbidden sign.

Finally, there is a prevalent stereotyping of Chinese tourists exhibiting transgressive behavior at the tulip fields partly generated by media reports. While there may have been reported cases, the current study demonstrates that unless certain moral disengagement mechanisms are at play, a majority of the tourists stick to the adherence of rules and regulations and do not necessarily exhibit transgressive behavior, especially at the tulip fields. Further, "moral actions are the product of the reciprocal interplay of cognitive, affective and social influences" (Bandura, 2002, p. 277).

Therefore, it can be said that social realities can influence moral action in some sense. In China, most farmers not wanting outsiders to enter their field use various conservative tools and resources to protect their land. It is a very common practice and hence the Chinese tourists somehow get used to passive perception. This results in a propensity to behave likewise at the fields, depending on the presence of conservative physical barriers. But in the Netherlands, the interviews conducted revealed that it is usually not a common practice to set up any protective fences for fields reflecting a comparatively less conservative and free mindset. Of course, there are tulip field farmers that have resorted to fencing in recent times, but this has mostly been an outcome of excessive tourists entering and damaging the fields, and such spatial protective strategies are still less in number compared to farming lands in China.

Local perspectives and tourism managerial strategies

The primary interviews and group discussions revealed six strategies for managing tourists at the tulip fields.

Spatial strategies such as fencing or creating physical boundaries

The adoption of physical barriers such as fences by the farmers is meant to reduce the negative impact of tourism on the tulip fields. In tourism destination management, the Doxey Irritation Index model is designed to represent the residents' anger level to cultural, social, environmental, and economic impact. This model divides the anger (or the lack of it) of residents into four levels: euphoria, apathy, irritation, and antagonism (Doxey, 1975). The putting up of a fence is indicative of the level of irritation among Dutch tulip farmers toward tourists. However, due to the COVID-19 pandemic, it is unclear where the farmers' attitudes toward tourists currently stand, on this irritation index. According to the Coordinator of the Flower Science Organization:

> *On crowded weekends, the farmers were continuously busy chasing people off their land. They could never rest, they had to work during workdays and chase people on weekends. They were very agitated about it and very tired.*

She added that it was quite challenging to convince farmers to install a fence during the initial stage because, with a fence in place, it is not convenient for a tractor to enter the field for harvesting purposes. The organization often lent barriers and fences to the farmers to ensure that tourists are aware of the field being private property. From April 2019, farmers started to install fences in their fields to keep the tourists out.

Arranging signboards at the fields

A signboard's purpose is to convey and communicate messages to tourists to ensure that the tourists behave properly. A signboard has multiple functions: it offers service information, directions about the route, and warnings as to what the tourists are forbidden to do. 'Enjoy the flowers, Respect our pride', the slogan board designed by the Flower Science Organization, tells the tourists that they should be taking selfies without trampling on the tulips. An entry-forbidden sign might be more effective still since it delivers a clearer message.

Arranging signboards in the fields could remind the tourists not to step in the fields, especially if there are several signs along the entire tulip route; it could help spread awareness that the fields are private. During the interview, the interviewees mentioned that the tourists could not distinguish the private and touristic areas since there is no clear signboard information to that effect—no wonder then that the tourists believe that the fields are open for entry. Language and effective communication through signboards is, therefore, an essential component in regulating tourist behavior. Moreover, signboards could regulate tourist behavior under circumstances of moral disengagement and psychologically inhibit impulses to misbehave.

Volunteering

Volunteer service has always been the foundation of operating public activities in the Netherlands. Volunteers are in a position to direct and manage the tourists' behavior when they intend to enter the field. When signboards are set up in the fields, it is necessary to have volunteers explain the information correctly to tourists. The group discussions with local stakeholders revealed that many tourists understood the information and behaved appropriately in the presence of volunteers. Volunteers also contribute to the tourists' sense of place. Tourists could have authentic experiences through communication with local volunteers.

> *[W]e want to have real people with a real story about farming and telling tourists why they should respect the fields. So, it is not some officer or tour guide who is providing you information, it's a person who has a brother or a neighbor or uncle who is a farmer and that is why he can speak from his heart.*
>
> [Coordinator of Flower Science]

Besides local people, there are volunteer 'ambassadors', who play an educational role by sharing knowledge and answering questions for tourists. More and more students have joined their team. The strategy of volunteering can be seen as a potential sustainable means of co-creation as well. Voluntourism, which is the merging of volunteering and tourism with a growing trend of tourists traveling to do some good in other communities, can be encouraged in tulip growing areas to bridge the gap between tourists and local farmers and contribute to knowledge dissemination about tulip farming practices and tourism. Such meaningful engagement between tourists, volunteer tourists, and locals could be useful for protecting the private tulip fields and fostering a sense of place.

Tourist dispersal

Cities such as Amsterdam and Den Haag have for a long time suffered from overtourism. However, other provinces, which also have tulip fields, get less attention from the tourists due to lack of promotion or marketing such has been the case with the tulip fields in the northern part of the Netherlands. In the long-term view, not all the provinces and residents can benefit from the tourism industry. Therefore, controlling the number of tourists by changing the marketing strategy is one of the solutions to reduce the damage to tulip fields and minimize the negative impact of overtourism.

In 2019, the Dutch government invested in rebranding the tourism strategy as 'NL Netherlands' rather than 'Holland' in the global market. Holland is just the name of one province that includes popular and well-marketed cities. To promote a more inclusive national identity, the authorities decided to use 'the Netherlands' on a logo to promote the entire country as a destination. The old logo, a tulip combined with 'Holland', contains a visual image that implies tourism related to tulip fields and souvenirs. The new logo is designed abstractly like a tulip, but it is straightforwardly applicable for all kinds of purposes such as in education, trade, and foreign embassy (Adams, 2019). The new strategy is to promote sustainable tourism and spread tourists to other places beyond Amsterdam. The tulip plays a vital role in representing the national identity, and the logo influences how tourists form their mental image of the Netherlands. The strategy of changing the logo is a symbolic movement of attaining maturity; it means that the focus of tourism in the Netherlands is shifting from tourism marketing to tourism management.

Setting up campaigns

The campaign 'Enjoy the flowers, Respect our pride' was established in February 2019 to stop tourists from trampling over tulips for taking photographs. Initially, the campaign covered the Bollenstreek area; later on, it spread online with further promotion by the NBTC. Gradually, more tulip growers in other regions joined the campaign. According to the NBTC official:

> *The campaign has several marketing instruments such as banners, websites, and flyers for explaining to visitors about the tulip fields. We should educate national but especially international visitors on how to approach our fields and to take care of the flowers.*

The Coordinator of Flower Science mentioned that the campaigns have some benefit as tourists were more careful with the tulips compared to the previous years. Besides this campaign, NBTC also has promoted 'tulip-friendly selfie' and 'watch your feet' hashtags for pictures and videos online. These campaigns also include the address of the experimental fields that allow tourists to enter and therefore help minimize the negative impacts of tourism on the flower fields. Moreover, this educational strategy aims to ensure that tourists conduct themselves in a morally acceptable manner. However, these campaigns started in 2019, and therefore, it is unclear to what extent their messages can regulate tourist behavior.

Digital technologies and smart apps

While volunteer services help to manage tourist behavior, campaigns have shortcomings due to over-reliance on volunteers. When tourists download a tulip route app online, they receive warning signs and instructions in advance. Therefore, there is less likelihood of them entering the restricted fields. With updated technology, designing a tulip route map could improve the tourists' experience. The best way to enjoy the tulip fields is to bike or drive around them. The Bollenstreek website has, accordingly, designed a smart app that offers guiding information for a bike, scooter, walking, and driving tours with the information presented in the picture, audio, and video format. Consequently, the tourists can check the map for directions and listen to information about the tulip fields (e.g., Holland Audio Guides—Flower Tour 2020). The same applies to Flevoland, where they have also invested in a tulip route app for the Flevoland province, with the route being open only during the tulip season. Overall, these smart apps cannot regulate tourist behavior as is done by the volunteers but can distribute the numbers of tourists during the peak season in particular hot-spot areas.

Conclusion

The understanding of transgressive behavior among tourists is relevant for the implementation of destination management strategies. The perception of transgressive behavior by Chinese tourists can be understood as a mechanism of moral disengagement that is likely to happen when individual tourists are affected by the context of collective behavior, which influences the individual moral standards—standards that one normally consults before taking action, and which reference the anticipated consequences of one's actions.

Additionally, the COVID-19 pandemic has forced us to re-think the damages done by mass tourism, brainstorm solutions for the future, and re-boot the tourism industry in line with the principles of sustainability. While efforts such as removing the orange tulip logo, rebranding tourism marketing, and Amsterdam's Enjoy and Respect' campaign imposing fines might help reduce the massive number of tourists visiting the tulip fields (once the pandemic is over), there is a need for introducing long-term, sustainable measures that strike a balance between tourists enjoying the tulip fields as well as refraining from damaging them for obtaining the perfect photo. Besides volunteering and promoting activities that actively raise tourist awareness and environmental consciousness toward tulip farming and fields, other value co-creation-based activities need to be organized that go beyond the tourist gaze, provide an insight into the Dutch local life, promote an authentic and emotional investment among the tourists, and establish more meaningful relationships between tourists and hosts.

In terms of study limitations, it remains a possibility that different demographic groups might have a different moral standard for transgressive behavior. For example, it would be interesting to analyze how people of different age groups and gender among Chinese tourists behave. Second, the study could have benefited from a larger sample size comprising local farmers and their perspectives on tourist behavior. While the current semi-structured interviews did not mention any instances of tourist misbehavior explicitly, it is possible that local narratives could have been different. Third, the first author's identity and positionality as a researcher of Chinese origin could have been an influencing factor in the interview responses and could have led to more guarded answers for the fear of not appearing to sound racist against Chinese tourists. When the COVID-19 outbreak began in China, there were negative and alarmed responses toward Chinese tourists in the international media. It is, therefore, unclear how Chinese tourists' travel desires and motivations will be influenced by the pandemic and how tourist-host interactions and management will undergo a transformation in the years to come.

References

Adams, C. (2019, December 27). Netherlands to drop 'Holland' as nickname. *Independent.* Retrieved August 13, 2020, from https://www.independent.co.uk/travel/news-and-advice/netherlands-holland-dutch-tourism-board-logo-a9261266.html

Bandura, A. (1991). Social cognitive theory of moral thought and action. In W. M. Kurtines & J. L. Gewirtz (Eds.), *Handbook of moral behavior and development: Theory, research and applications* (pp. 71–129). Hillsdale, NJ: Erlbaum.

Bandura, A. (1999). Moral disengagement in the perpetration of inhumanities. *Personality and Social Psychology Review, 3*(3), 193–209. doi:10.1207/s15327957pspr0303_3

Bandura, A. (2002). Selective moral disengagement in the exercise of moral agency. *Journal of Moral Education, 31*(2), 101–119. doi:10.1080/0305724022014322

Bandura, A., Barbaranelli, C., Caprara, G. V., & Pastorelli, C. (1996). Mechanisms of moral disengagement in the exercise of moral agency. *Journal of Personality and Social Psychology, 71*(2), 364–374. doi:10.1037/0022–3514.71.2.364

Doxey, G. V. (1975). Theory of visitor-resident irritants, methodology and research inferences: The impact of tourism. In *Sixth Annual Conference of the Travel Research Association.*

Guitard, P., Ferland, F., & Dutil, É. (2005). Toward a better understanding of playfulness in adults. *OTJR: Occupation, Participation and Health, 25*(1), 9–22. doi:10.1177/153944920502500103

Heylighen, F. (1992). A cognitive-systemic reconstruction of Maslow's theory of self-actualization. *Behavioral Science, 37*(1), 39–58. doi:10.1002/bs.3830370105

Kang, M., & Moscardo, G. (2006). Exploring cross-cultural differences in attitudes towards responsible tourist behaviour: A comparison of Korean, British and Australian tourists. *Asia Pacific Journal of Tourism Research, 11*(4), 303–320. doi:10.1080/10941660600931143

Kuo, N.-T., Chang, K.-C., Cheng, Y.-S., & Lin, J.-C. (2016). Effects of tour guide interpretation and tourist satisfaction on destination loyalty in Taiwan's Kinmen battlefield tourism: Perceived playfulness and perceived flow as moderators. *Journal of Travel & Tourism Marketing, 33*(supl), 103–122. doi:10.1080/10548408.2015.1008670

Li, M., Sharpley, R., & Gammon, S. (2019). Towards an understanding of Chinese tourist photography: Evidence from the UK. *Current Issues in Tourism, 22*(5), 505–521. doi:10.1080/13683500.2017.1377690

Maslow, A. H. (1943). A theory of human motivation. *Psychological Review, 50*(4), 370–396. doi:10.1037/h0054346

Sharma, N. (2020). Dark tourism and moral disengagement in liminal spaces. *Tourism Geographies, 22*, 1–25. doi:10.1080/14616688.2020.1713877

Tolkach, D., Pratt, S., & Zeng, C. Y. H. (2017). Ethics of Chinese & Western tourists in Hong Kong. *Annals of Tourism Research, 63*, 83–96. doi:10.1016/j.annals.2017.01.004

Urry, J. (1990). The consumption of tourism. *Sociology, 24*(1), 23–35.

Urry, J., & Larsen, J. (2011). *The tourist gaze 3.0.* London: Sage.

Wang, N. (1999). Rethinking authenticity in tourism experience. *Annals of Tourism Research, 26*(2), 349–370. doi:10.1016/s0160–7383(98)00103-0

13 The future of Chinese outbound tourism

An Australian perspective

Clare Lade and Elspeth Frew

Introduction

The Chinese outbound tourism market reached almost 155 million during 2019 and is expected to be worth US$365 billion for the 2019–2025 period (PR Newswire, 2019; Statista, 2020a). Substantial growth in the Chinese outbound market, influenced largely by societal, economic and technological changes in China, creates a range of implications for many receiving countries (Lade et al., 2020; Yu, 2014). With Australia being one of the first western countries granted Approved Destination Status (ADS), this enabled Tourism Australia to directly market to Chinese consumers (Australian Trade and Investment Commission [Austrade], 2020). A five-year memorandum of understanding (MOU) was signed between China National Tourism Administration and the Australian Government on the ADS scheme in 2006 and then renewed in 2016 (Austrade, 2020). Subsequently, China became one of the key inbound markets for Australia, and its forecasted growth presented many future tourism opportunities (Pham et al., 2017).

Prior to the Australian government introduction of travel bans on international arrivals in March 2020, in 2019, China was Australia's largest inbound market for both visitor arrivals and total spend (Tourism Australia, 2020a), generating US$12 billion from 1.2 million visitors each year (Ludlow, 2020). The Chinese traveller to Australia on average stayed 45 nights and spent US$9,418 with 51% travelled for the purpose of a holiday, with 58% being return visitors (Tourism Australia, 2020b). Additionally, Australian tertiary education providers were largely reliant upon international students from China with approximately 160,000 Chinese students enrolled in the higher education (Statista, 2020b). In 2018/2019, Chinese students to Australian tertiary sector dominated Australian academic exports, representing a contribution of approximately US$12.1 billion (Verrender, 2020).

In the state of Victoria, many Chinese travellers come to visit family and friends in Australia's large local Chinese communities, undertake education, and visit the variety of nature-based wonders such as the Great Ocean Road and historical attractions such as Sovereign Hill (Joyce, 2015). This chapter examines the future of Chinese outbound tourism from an

DOI: 10.4324/9781003121329-15

Australian perspective. Considerations including economic factors, tourist desires, needs and travel experiences, meeting demand, host-visitor relations and the impact of the COVID-19 pandemic on future outbound tourism to Australia will be explored.

Economic factors

Over the last 20 years, China has experienced increasing disposal income due to an economic boom, resulting in many Chinese travelling abroad (Lin et al., 2015). When compared to those traditional profitable tourist markets such as the United Kingdom, USA, Germany, France, and the Nordic countries, China demonstrates a reasonably high economic growth rate however a comparatively lower GDP per capita (Xie & Tveterås, 2020). This economic situation suggests that a larger share of the population reaches a level of purchasing power which enables them to travel overseas, therefore resulting in a more rapid growth in outbound travel. With a larger income elasticity, as estimated by Lin et al. (2015), outbound travel may be viewed as a luxury good which generally responds well to income growth among potential travellers.

Prior to COVID-19, it was predicted that it was likely the number of people travelling outbound from China would remain considerable even in the event of a major economic downturn negatively impacting travel affordability for an extended period of time (Global Wellness Institute [GWI], 2019). At 155 million travellers, China's outbound tourism rate is now above the USA (84 million), almost twice that of the United Kingdom (70.4 million) and five times larger than that of France (27 million) (Statista, 2020a; United Nations World Tourism Organization [UNWTO], 2018). The social, economic and technological factors that have contributed to this significant growth of the China outbound travel market include the 'political liberalization and social tolerance for diversified sociocultural values' (Cai et al., 2008; Xie & Tveterås, 2020, p. 112); changes to the visa issuing process for Chinese travellers (Wang & Davidson, 2010); and, investment in the areas of innovation and digitisation (Kang, 2020). Quality air connectivity (Tveteras & Roll, 2014), price competitiveness and exchange rates (Xie & Tveterås, 2020) are also key influencing factors.

Outbound Chinese tourism and consumption trends

The characteristics of the Chinese traveller are changing and evolving more quickly than any other international market to Australia (Destination NSW, 2015). Being able to engage in a meaningful experience is the foremost priority for Chinese tourists when selecting an overseas travel destination or tourist attractions with this factor considered more important than the costs incurred (Nielsen, 2017; Xie & Tveterås, 2020). Chinese outbound tourists expressed that beauty and uniqueness at the destination (56%); visitation to theme parks (45%); ease of visa procedures (45%); and being made to feel welcome at the destination (35%) were important to them. Forty-five percent of Chinese travellers also stated that their safety at a given destination would affect their travel choice (Nielsen, 2017; Zheng & Zou, 2016; Zou & Meng, 2019). Media travel warnings, language barriers, lack of information, culture and visa regulations (Li, Lai et al., 2011; Sparks & Pan, 2009; Wu, 2015) were identified as being the main barriers to Chinese outbound travel.

More specifically in relation to travel to Australia, Chinese tourists identified the top attractions as being the Sydney Opera House and Harbour Bridge, the Gold Coast and Great Barrier Reef in Queensland. In terms of travel expenditure, Chinese travellers spent the most on shopping, followed by accommodation and dining with their location spend averaging US$3,541 p/p in Australia, third only to Europe (US$3,754) and USA (US$4,462) (China Tourism Academy, 2018; Nielsen, 2017). It must be recognised that the shopping expenditure may not indeed be for the individual tourists themselves, with Lin (2016) suggesting that gifts constituted over 80% of travel-related expenditure by Chinese tourists. Factors affecting Chinese shopping habits included discounts (41%), payment method (42%) and price (40%) as well as quality, location and the merchants' hospitality (Nielsen, 2017).

Value for money was also considered important (Li, Zhang et al., 2011). The duty-free shops and large supermarkets were identified as the most popular outlets in Australia for Chinese travellers with the most popular items purchased being related to skin care and beauty, clothing (including hats and shoes), local specialities and souvenirs (Nielsen, 2017). In addition, travelling with SMART communication devices has become a new outbound trend. Sixty-five percent of Chinese tourists used mobile payment platforms during overseas travel (compared to 11% of non-Chinese tourists) which demonstrates the importance for Australian businesses and the tourism industry to cater to these tourist's spending needs, particularly in regards to provision of suitable shopping, attractions and accommodation (Nielsen, 2017). Given the strong demand among Chinese citizens for overseas travel, it is important to understand the types of holiday which is of interest to Chinese people. The following discussion highlights the importance of health and wellness trends and the importance of safety aspects.

Health and wellness trends

The concept of health and wellness is complex. No universal definition of health tourism exists, and often, the term 'health' is used to encompass all health-related tourist activities (Connell, 2006; Gonzales et al., 2001; Hall, 2011; Mueller & Kaufmann, 2001; Voigt et al., 2010), including medical and holistic activities. Wellness is often regarded as a subset of health tourism and is based on the notion of improving and maintaining health conditions

(Gonzales et al., 2001; Voigt et al., 2010). The wellness concept typically concerns spa, massage, beauty, health camps, diet, physical exercise, ecotourism, pilgrimage, sport and related retreats. Smith and Puczko (2015, p. 207) suggest that wellness tourism extends beyond spa tourism to also encompass 'healthy cuisine, specific fitness or body-mind-spirit regimes, active ageing or longevity programmes, learning, adventure, spiritual enlightenment and personal growth, all of which take place in purpose-built centres'. Many Asian health and wellness-related practices are derived from Eastern religious beliefs, for example, Buddhist elements within meditation, South Asian yoga and Ayurvedic practices (Laing & Weiler, 2008).

In China, the wellness concept is more prominent than that of health with Yangsheng (Chinese word for wellness) used throughout Chinese history and permeating Chinese everyday life (Huang & Xu, 2014). Longer life expectancies together with falling birth rates have contributed to the world's ageing population growth (Lade et al., 2020; World Health Organisation [WHO], 2011). The United Nations (UN) (2010) estimated that by 2020, the number of people living in China over the age of 60 years would be 248 million, with this figure increasing to 437 million by the year 2050. Yang Honghao, an expert with the China Tourism Academy, acknowledges that "in the past, wellness tourism has been related to older citizens and their lifestyles. But now, more and more young Chinese favour in-depth travels for the purpose of health preservation" (Xinhua, 2018). During 2017, 70.2 million wellness-related trips were made in China with many Chinese seeking healthcare and medical treatments abroad due to a lack of quality medical care being available domestically (GWI, 2017).

It is estimated that half a million outbound Chinese medical travellers spend at least US$10 billion a year, creating profitable demand in other countries for related health and medical services (Global Wellness Summit [GWS], 2019). The services which these Chinese medical travellers are seeking range from minor treatments including health check-ups, prescription fillings and cosmetic and surgical procedures to more advanced treatments such as cardiology, oncology and neurology programmes, fertility treatment, obstetrics, gynaecology and joint replacement (GWI, 2017). Over-the-counter medicines and health supplements are also high in demand. With China being the world's fastest-growing wellness tourism market, having increased from the 11th to the 3rd position in wellness tourism globally in a relatively short amount of time and with anticipated substantial growth ahead, much opportunity exists for the Australian health and wellness sector (GWI, 2017).

A number of forest bathing and tree stay experiences are offered in the state of Victoria currently where tourists are encouraged to unplug and destress and to engage in these experiences (Visit Victoria, 2020). Additionally, Tourism Australia (2020c) highlights wellness-inspired road trips in Australia, including day spas on the Gold Coast and health and wellness retreats located all around Australia (including Gwinganna Lifestyle Retreat on the Gold Coast, Gaia Retreat and Spa in Byron Bay, Daintree Eco Lodge and Spa, Hepburn Bathhouse and Spa, Peninsula Spa and Hot Springs, Kangaroo Island Health Retreat, Injidup Spa Retreat Margaret River and Harmony Harvest Retreat in Hobart).

Many of these Australian day spas and wellness retreats are situated in regional locations surrounded by diverse natural environment, and while forest and nature therapy are a relatively new concept in Australia, potential does exist to expand within this area in order to attract Chinese outbound tourists. Health and wellness in China will only continue to grow from its traditional focus on acupuncture, herbal medicine and hot springs, to encompass modern wellness programs focusing on stress reduction, mental health well-being and overall good health (Heung & Kucukusta, 2012; Lade et al., 2020). Chinese tourists are likely to seek out and incorporate these wellness-related activities into their outbound travel, and therefore destinations within key travel markets, such as Australia, need to ensure that such high-quality experiences are offered.

Consideration of safety aspects

Following the September 11, 2001 attacks, travel safety began to attract growing interest globally (Zheng & Zou, 2016). Four major risk factors which have been identified as influencing tourists' safety perceptions include terrorism, war and political instability, crime and health hazards (Adam & Adongo, 2016; Barker et al., 2003; Botterill et al., 2013; Cartwright, 2000; Coshall, 2003; Neumayer, 2004; Samitas et al., 2018; Weaver, 2000). While all of these factors may potentially deter tourists from visiting particular destinations, political instability, terrorism acts and war were identified as those imposing a greater influence within the holiday planning stage (Cavlek, 2002).

Chinese culture traditionally reflects caution surrounding safety and well-being issues (Zou & Zheng, 2012). The Chinese consider destination safety during the different travel stages, especially by preparing fully before travelling and paying attention to potential safety issues which may arise during their trips. Jin et al. (2019) explored the impact of five major crisis events affecting Chinese outbound tourism to Japan and South Korea between 2005 and 2017, determining that political crisis events at a national level demonstrated a greater impact than natural disasters, epidemic disease and financial crisis. During the severe acute respiratory syndrome (SARS) outbreak, many Chinese people chose to not travel; however, the tourism industry recovered as soon as the outbreak was controlled (Han, 2020).

The way that Chinese tourists travel however might change following the COVID-19 pandemic (Han, 2020). The global COVID-19 pandemic has been more globally widespread than the SARS outbreak during 2003, infecting 68.8 million people in 218 countries and resulting in 1.57 million deaths to date compared to 8,098 reported SARS cases in 30 countries, of which 774 deaths resulted (WHO, 2020). Chinese outbound tourists "will go overseas if their safety and health can be ensured" and may search for more meaningful trips, be more inclined to travel with family and seek out more natural and cultural experiences (Han, 2020). The Australian Government's strict response to the COVID-19 pandemic, resulting in minimal active cases and deaths relative to others worldwide, may prove Australia to be a more popular travel destination than other countries for the Chinese once travel restrictions are lifted.

Impact of COVID-19 on Chinese tourists to Australia

Since the emergence of COVID-19, some countries may appear less attractive to the Chinese outbound travel market due to the high prevalence of cases of COVID-19 such as in the United States (Chung, 2020). Australia may be perceived as a more attractive destination given the relatively lower levels of COVID-19 cases experienced during 2020 and the associated strict quarantine measures and testing regime. However, the Australian Government prohibited Chinese nationals from entering Australia from February 1, 2020 if they had been in China within 14 days (Francis, 2020). Then, in March 2020, a travel ban was placed on all international arrivals into Australia with only returning Australian citizens and permanent residents allowed to enter the country. This resulted in June 2020, of having China's tourist arrivals to Australia down by 37% and expenditure down by 29% in comparison to the same time the previous year (Tourism Australia, 2020b).

As an illustration of the impact of COVID-19 on the Australian tourism industry, the following discussion highlights tourism in the state of Victoria to reflect the strength and importance of the Chinese market. The State of Victoria is an attractive destination for Chinese tourists, and the recognition of the potential of and benefit from Chinese visitors was reflected by the state government's development of the 2016 Victoria China Strategy, which reinforced the desire to support and encourage more Chinese tourists to Victoria. In 2017, the largest share of visitor expenditure in Victoria was from China, and this represented 621,600 visitors. This number of visitors represented the most to the state from any other country and was more than the next six markets combined. The 2017 figures reflected a 27% increase in Chinese visitors since 2015 and was measured as outpacing the growth of all international visitors to Victoria in the same period (Victorian State Government, 2018a).

In 2018, the Victorian State Government launched a marketing campaign aimed at attracting young sophisticated, independent Chinese millennials who were looking for local and authentic Australian experiences. The campaign was aimed at residents of cities with already established direct flights to Melbourne such as Shanghai, Guangzhou, Beijing and Shenzhen. The campaign utilised the young Chinese actor Li Xian to highlight Victoria's attractions and focused on the laneways of Melbourne, the 12 Apostles, the Yarra Valley and showcased Melbourne's local produce, major sporting and cultural events, unique wildlife and natural landscapes (Victorian State Government, 2018b).

Many regional tourist attractions relied heavily on the Chinese market. For example, a pick-your-own berry farm called Berry World in the regional town of Timboon relied on tourism for about 45% of their income and during peak periods, nearly half of the farm's visitors were Chinese visitors (Francis, 2020). The regional Victoria Phillip Island Nature Park, which includes the world-famous Penguin Parade, usually welcomes more than 275,000 Chinese visitors a year to the site, but the ban on Chinese visitors has had a heavy financial impact (Francis, 2020). Similarly, Sovereign Hill, the regionally located gold mining historic village, was traditionally very popular among Chinese visitors with an estimated 10% of the 700,000 annual visitors being from China. This popularity reflected that the site is an easy one-hour drive from Melbourne and reflects that to ensure that the Chinese visitors had a memorable Australian experience, the owners worked closely with the inbound Chinese wholesalers and tour operators. In addition, over the last 20 years, the owners introduced guided tours by Chinese-speaking staff, and they provided visitors with the opportunity to discover their ancestor's role in the 19th-century gold rush.

These examples serve to highlight the reliance placed on the Chinese market among Australian tourism businesses and the devastating impact that the coronavirus bans have had on their businesses. However, even prior to the COVID-19 situation, there were concerns that the growing influx of Chinese tourists could slow if Australia became complacent, in such areas as infrastructure and the missed opportunities in regards to attracting return visits by providing new and exciting products (Fang, 2019).

Impact of the Chinese-Australian trade and diplomatic situation on Chinese outbound travel to Australia

Diplomatic and trade relations between Australia and China spiralled downwards in the wake of the Australian government call for an independent international inquiry into the origins of COVID-19 (Henderson et al., 2020) and the criticism of Beijing's treatment of the people of Hong Kong (Tang, 2020). However, the economic and trade relationship between Australia and China was problematic for several years prior to COVID-19, particularly in relation to the purchase of Australian iron ore, with Australia supplying 65% of all China's iron ore. Iron ore has been described as China's economic lifeblood as it is a key ingredient in the manufacturing of steel which is used on Chinese stimulus-related infrastructure spending to build residential and office towers and high-speed rail. Since mid-2020, iron ore prices increased significantly reflecting China's increased demand for steel production and problems in the supply of iron ore from Brazil and Guinea, West Africa (Verrender, 2020).

In an apparent response to the increase in the price and supply of iron ore and the Australian government COVID-19 comments, the deteriorating relationship between China and Australia has resulted in the Chinese government placing tariffs and bans on Australian barley, wine and rock lobster and an unofficial ban on Australian coal. A ban on Australia exporting coal to China could have significant economic implications as Australia in 2019 exported coal worth almost US$14 billion to China (Henderson et al., 2020). In turn, Beijing has accused Australia of dumping grain and wine on the Chinese market and unfairly subsidising farmers (Verrender, 2020). Some commentators have suggested that it may take years to recover from this diplomatic situation with any chance of mending the relationship seeming to be out of reach (Verrender, 2020).

It is not known what the long-term impact may have on Chinese tourism to Australia once the COVID-19 travel ban has been lifted, but it is suggested that a normal return to the massive China tourism trade or education industry is unlikely. One commentator suggested, "it is not that ordinary Chinese citizens won't want to come. It is just that they will be discouraged and it is unlikely they will ever arrive in the same numbers" (Verrender, 2020). One way such discouragement could occur might be for the Chinese authorities to highlight a perceived Australian racism against Chinese visitors to Australia, similar to the claim made by the Chinese Ministry of Culture and Tourism in June 2020 when they issued an alert warning against travel to Australia, citing a "significant increase" in racist attacks on "Chinese and Asian people" (Walden, 2020).

Conclusion

The Chinese outbound tourism has grown significantly over the last 20 years reflecting the growing importance of the Chinese economy and the development of a large Chinese middle class. Chinese tourism to Australia grew significantly over this same period, and Australia received a steady growth in tourists from China prior to the COVID-19-related travel ban implemented in March 2020.

Chinese tourists to Australia are interested in a variety of tourist experiences ranging from nature-based tourism to health and wellness tourism. They enjoy the opportunity to purchase souvenirs and feel comfortable when their safety concerns are addressed. Australian tourism and hospitality businesses have developed and adapted their products and services accordingly in order to satisfy the Chinese tourist needs. However, recent trade and diplomatic tensions between the Chinese and Australian governments are likely to have an impact on the future visitation of outbound Chinese tourists to Australia with it being expected take a few years to return to those pre-2020 Chinese tourist visitor numbers.

References

Adam, I., & Adongo, C. A. (2016). Do backpackers suffer crime? An empirical investigation of crime perpetrated against backpackers in Ghana. *Journal of Hospitality and Tourism Management*, *27*, 60–67. doi:10.1016/j.jhtm.2016.03.003

Australian Trade and Investment Commission. (2020). *Working with China*. https://www.austrade.gov.au/Australian/Tourism/Working-with-China/ADS

Barker, M., Page, S. J., & Meyer, D. (2003). Urban visitor perceptions of safety during a special event. *Journal of Travel Research*, *41*(4), 355–361. doi:10.1177/0047287503041004004

Botterill, D., Pointing, S., Hayes-Jonkers, C., Clough, A., Jones, T., & Rodriguez, C. (2013). Violence, backpackers, security and critical realism. *Annals of Tourism Research*, *42*, 311–333. doi:10.1016/j.annals.2013.01.007

Cai, L. A., Li, M., & Knutson, B. J. (2008). Research on China outbound market: A meta-review. *Journal of Hospitality & Leisure Marketing*, *16*(1–2), 5–20. doi:10.1080/10507050802096778

Cartwright, R. (2000). Reducing the health risks associated with travel. *Tourism Economics*, *6*(2), 159–167. doi:10.5367/000000000101297550

Cavlek, N. (2002). Tour operators and destination safety. *Annals of Tourism Research*, *29*(2), 478–496. doi:10.1016/s0160-7383(01)00067-6

China Tourism Academy. (2018). *Annual report of China outbound tourism development 2018*. Beijing: Tourism Education Press. http://eng.ctaweb.org.cn/ctaen/c24/leftd_list.shtml

Chung, N. (2020, July 20). US loses its shine with Chinese tourists amid COVID-19. *TTG Asia*. Retrieved December 13, 2020, from https://www.ttgasia.com/2020/07/20/us-loses-its-shine-with-chinese-tourists-amid-covid-19/

Connell, J. (2006). Medical tourism: Sea, sun, sand and … surgery. *Tourism Management*, *27*(6), 1093–1100. doi:10.1016/j.tourman.2005.11.005

Coshall, J. T. (2003). The Threat of terrorism as an intervention on international travel flows. *Journal of Travel Research*, *42*(1), 4–12. doi:10.1177/0047287503253901

Destination NSW. (2015). *China market toolkit: Information, resources, development, opportunities*. https://www.destinationnsw.com.au/wp-content/uploads/2015/06/China-Market-Toolkit.pdf

Fang, J. (2019, January 26). Chinese tourism boom could slow down if Australia becomes complacent, expert warns. *ABC News*. Retrieved January 5, 2021, from https://www.abc.net.au/news/2019-01-26/warnings-not-to-be-complacent-with-chinese-tourism-market/10746516

Francis, C. (2020, February 12). Regional tourism hit by China ban. *Weekly Times*. https://www.weeklytimesnow.com.au/subscribe/news/1/?sourceCode=WTWEB_WRE170_a_GGL&dest=https%3A%2F%2Fwww.weeklytimesnow.com.au%2Fnews%2Fvictoria%2Fcoronavirus-regional-victorian-tourism-hit-by-china-ban%2Fnews-story%2Fb490f236145bb4d8118d3cf284b813f8&memtype=anonymous&mode=premium

Global Wellness Institute. (2017). *Chinese outbound wellness tourism will be a huge force*. https://globalwellnessinstitute.org/global-wellness-institute-blog/2017/06/20/2017-6-20-monthly-barometer-wellness-editionchinese-outbound-wellness-tourism-will-be-a-huge-force/

Global Wellness Summit. (2019). *China – uncovering the wealth in wellness*. https://www.globalwellnesssummit.com/2019-global-wellness-trends/china-uncovering-wealth-in-wellness/

Gonzales, A., Brenzel, L., & Sancho, J. (2001). *Health tourism and related services: Caribbean development and international trade.* Final Report, 20.

Hall, C. M. (2011). Health and medical tourism: A kill or cure for global public health? *Tourism Review*, *66*(1/2), 4–15. doi:10.1108/16605371111127198

Han, Y. (2020, July 14). Optimism seen for recovery of outbound travel. *Global China Daily*. https://global.chinadaily.com.cn/a/202007/14/WS5f0d083da310834817259349.html

Henderson, A., Dziedzic, S., Oaten, J., & Patidar, S. (2020, December 14). China's the Global Times appears to confirm a ban on Australian coal imports amid perilous trade tensions. *ABC News Online*. Retrieved December, 13, 2020, from https://www.abc.net.au/news/2020-12-14/global-times-reports-australian-coal-exports-blocked-by-china/12983336

Heung, V. C. S., & Kucukusta, D. (2012). Wellness tourism in China: Resources, development and marketing. *International Journal of Tourism Research*, *15*(4), 346–359. doi:10.1002/jtr.1880

Huang, L., & Xu, H. (2014). A cultural perspective of health and wellness tourism in China. *Journal of China Tourism Research*, *10*(4), 493–510. doi:10.1080/19388160.2014.951752

Jin, X. (Cathy), Qu, M., & Bao, J. (2019). Impact of crisis events on Chinese outbound tourist flow: A framework for post-events growth. *Tourism Management*, *74*, 334–344. doi:10.1016/j.tourman.2019.04.011

Joyce, E. (2015, December 18). Good year for tourism but work still to be done. *Herald-Sun*. https://www.heraldsun.com.au/subscribe/news/1/?sourceCode=HSWEB_WRE170_a_GGL&dest=https%3A%2F%2Fwww.heraldsun.com.au%2Fbusiness%2Finternational-tourism-win-for-victoria-but-there-is-more-work-to-be-done%2Fnews-story%2F2740c009393f69b77ab3eeabb91867c0&memtype=anonymous&mode=premium&v21=dynamic-cold-test-score&V21spcbehaviour=append

Kang, X. (2020, November 10). Why China's economy keeps booming even after Covid. *The Diplomat*. Retrieved December, 13, 2020, from https://thediplomat.com/2020/11/why-chinas-economy-keeps-booming-even-after-covid-19/

Lade, C., Strickland, P., Frew, E., Willard, P., Nagpal, S., Cherro Osorio, S., & Vitartas, P. (2020). *International tourism futures: The drivers and impacts of change*. Oxford: Goodfellow Publishers.

Laing, J., & Weiler, B. (2008). Mind, body and spirit: Health and wellness tourism in Asia. In J. Cochrane (Ed.), *Asian tourism: Growth and change* (pp. 379–389). Oxford: Elsevier.

Li, M., Zhang, H., Mao, I., & Deng, C. (2011). Segmenting Chinese outbound tourists by perceived constraints. *Journal of Travel and Tourism Marketing*, *28*(6), 629–643. doi:10.1080/10548408.2011.598747

Li, X., Lai, C., Harrill, R., Kline, S., & Wang, L. (2011). When east meets west: An exploratory study on Chinese outbound tourists' travel expectations. *Tourism Management*, *32*(4), 741–749. doi:10.1016/j.tourman.2010.06.009

Lin, L. (2016). Food souvenirs as gifts: Tourist perspectives and their motivational basis in Chinese culture. *Journal of Tourism and Cultural Change*, *15*(5), 439–454. doi:10.1080/14766825.2016.1170841

Lin, V. S., Liu, A., & Song, H. (2015). Modeling and forecasting Chinese outbound tourism: An econometric approach. *Journal of Travel and Tourism Marketing*, *32*(1–2), 34–49. doi:10.1080/10548408. 2014.986011

Ludlow, M. (2020, June 26). Long road back to Chinese gold rush for Sovereign Hill. *The Australian Financial Review*. Retrieved December 13, 2020, from https://www.afr.com/companies/tourism/long-road-back-to-chinese-gold-rush-for-sovereign-hill-20200626-p556fg

Mueller, H., & Kaufmann, E. L. (2001). Wellness tourism: Market analysis of a special health tourism segment and implications for the hotel industry. *Journal of Vacation Marketing*, *7*(1), 5–17. doi:10.1177/135676670100700101

Neumayer, E. (2004). The impact of political violence on tourism: Dynamic cross-national estimation. *Journal of Conflict Resolution*, *48*(2), 259–281. doi:10.1177/0022002703262358

Nielsen. (2017). *Outbound Chinese tourism and consumption trends*. https://www.nielsen.com/wp-content/uploads/sites/3/2019/05/outbound-chinese-tourism-and-consumption-trends.pdf

Pham, T. D., Nghiem, S., & Dwyer, L. (2017). The determinants of Chinese visitors to Australia: A dynamic demand analysis. *Tourism Management*, *63*, 268–276. doi:10.1016/j.tourman.2017.06.015

PR Newswire. (2019, September 25). China outbound tourists visit and spending report 2019- Forecast to 2025 for the $365 Billion Industry. *PR Newswire*. Retrieved January 5, 2021, from https://www.prnewswire.com/news-releases/china-outbound-tourists-visit-and-spending-report-2019---forecast-to-2025-for-the-365-billion-industry-300925281.html

Samitas, A., Asteriou, D., Polyzos, S., & Kenourgios, D. (2018). Terrorist incidents and tourism demand: Evidence from Greece. *Tourism Management Perspectives*, *25*, 23–28. doi:10.1016/j.tmp.2017.10.005

Smith, M., & Puczkó, L. (2015). More than a special interest: Defining and determining the demand for health tourism. *Tourism Recreation Research*, *40*(2), 205–219. doi:10.1080/02508281.2015.1045364

Sparks, B., & Pan, G. W. (2009). Chinese outbound tourists: Understanding their attitudes, constraints and use of information sources. *Tourism Management*, *30*(4), 483–494. doi:10.1016/j.tourman.2008.10.014

Statista. (2020a). *Number of outbound tourists departing from China from 2010-2019*. https://www.statista.com/statistics/1068495/china-number-of-outbound-tourist-number/#:~:text=In%202019%2C%20the%20number%20of, from%20other%20countries%20in%202019

Statista. (2020b). *Number of Chinese students in Australia by education sector*. https://www.statista.com/statistics/430276/number-of-chinese-students-in-australia-by-education-sector/

Tang, D. (2020, December 18). Mass power cuts follow China's coal imports ban. *The Times*. https://www.thetimes.co.uk/article/mass-blackouts-after-china-cuts-australian-coal-imports-phx3fgvtg

Tourism Australia. (2020a). *China*. https://www.tourism.australia.com/en/markets-and-stats/market-regions/greater-china/china.html

Tourism Australia. (2020b). *International market performance statistics*. https://www.tourism.australia.com/en/markets-and-stats/tourism-statistics/international-market-performance.html

Tourism Australia. (2020c). *Health and wellness experiences*. https://www.australia.com/en/things-to-do/health-and-wellness-experiences.html

Tveteras, S., & Roll, K. H. (2014). Non-stop flights and tourist arrivals. *Tourism Economics*, *20*(1), 5–20. doi:10.5367/te.2013.0263

United Nations. (2010). *World population ageing 2009*. https://www.un.org/en/development/desa/publications/world-population-ageing-2009.html

United Nations World Tourism Organization. (2018). *Country profile outbound tourism*. https://www.unwto.org/country-profile-outbound-tourism

Verrender, I. (2020, December 14). As Australia's relationship with China deteriorates beyond repair, we need to find new trade partners. *ABC News Online*. Retrieved January 4, 2021, from https://www.abc.net.au/news/2020-12-14/australia-has-to-look-beyond-china-as-relationship-breaks-down/12979900

Victorian State Government. (2018a, May). *Victoria's China strategy progress report 2018*. Victorian State Government. https://www.vic.gov.au/victorias-china-strategy

Victorian State Government. (2018b). *New China campaign to drive Victorian tourism and jobs*. https://www.premier.vic.gov.au/new-china-campaign-to-drive-victorian-tourism-and-jobs-0

Visit Victoria. (2020). *Forest bathing*. Retrieved January 5, 2021, from https://www.visitvictoria.com/Features/Forest-bathing

Voigt, C., Laing, J., Wray, M., Brown, G., Howat, G., Weiler, B., & Trembath, R. (2010). *Health tourism in Australia: Supply, demand and opportunities*. Gold Coast: CRC for Sustainable Tourism Pty Ltd.

Walden, M. (2020, June 8). Australia says China travel warning 'unhelpful' amid escalating diplomatic row. *ABC News*. Retrieved December 13, 2020, from https://www.abc.net.au/news/2020-06-08/china-coronavirus-travel-ban-australia-diplomatic-row/12332016

Wang, Y., & Davidson, M. C. G. (2010). Chinese holiday makers' expenditure: Implications for marketing and management. *Journal of Hospitality Marketing and Management*, *19*(4), 373–396. doi:10.1080/19368621003667101

Weaver, D. B. (2000). The exploratory war-distorted destination life cycle. *International Journal of Tourism Research*, *2*(3), 151–161. doi:10.1002/(sici)1522-1970(200005/06)2:3<151::aid-jtr196>3.0.co;2-g

World Health Organisation. (2011). *Global health and ageing*. National Institute on Ageing National Institutes of Health. https://www.who.int/ageing/publications/global_health.pdf

World Health Organisation. (2020). *WHO coronavirus disease (Covid-19) dashboard*. https://covid19.who.int/

Wu, M. Y. (2015). Driving an unfamiliar vehicle in an unfamiliar country: Exploring Chinese recreational vehicle tourists' safety concerns and coping techniques in Australia. *Journal of Travel Research*, *54*(6), 801–813. doi:10.1177/0047287514532364

Xie, J., & Tveterås, S. (2020). Economic drivers for the Chinese tourists. *Scandinavian Journal of Hospitality and Tourism*, *20*(2), 110–125. doi:10.1080/15022250.2020.1734076

Xinhua. (2018, October 4). Across China: Wellness tourism gains popularity among young white collars. *Xinhuanet*. Retrieved December 16, 2020, from http://www.xinhuanet.com/english/2018-10/04/c_137511477.htm

Yu, L. (2014). *Consumption in China: How China's new consumer ideology is shaping the nation*. New York: John Wiley.

Zheng, X. M., & Zou, Y. G. (2016). Safety and security. In J. Jafari & H. Xiao (Eds.), *Encyclopedia of tourism* (pp. 814–815). Cham: Springer International.

Zou, Y., & Meng, F. (2019). Chinese tourists' sense of safety: Perceptions of expected and experienced destination safety. *Current Issues in Tourism*, *23*, 1–14. doi:10.1080/13683500.2019.1681382

Zou, Y. G., & Zheng, X. M. (2012). An empirical study on the affecting factors of tourists' sense of security in tourism destinations. *Tourism Tribune*, *27*(1), 49–57.

Part III

Chinese outbound tourists' shopping and health

14 The shopping behaviour of Chinese outbound tourists

Lina Zhong, Liyu Yang and Alastair M. Morrison

Introduction

The tourism industry has undergone major modifications due to the technological development, social, and cultural changes in the world (Cohen & Cohen, 2012). Convenient transportation, visa application procedures, and the development of information technology have made international exchanges more and more frequent, and the scale of the outbound tourism market has been expanding year by year. The relaxation of tourism policies and visa issuance has greatly promoted the growth of tourism (Wang & Davidson, 2010). In 2016, there are 122 million Chinese outbound tourists, and the expenditure was 109.8 billion dollars (China Tourism Academy, 2017), making it the number one in the world (Xu et al., 2018). In 2019, China's outbound tourism market reached 155 million, and outbound tourists spend more than 133.8 billion dollars (China Tourism Academy, 2020). The booming of China's outbound tourism is playing an important role in boosting the economy of other countries (Xu et al., 2018).

Shopping is one of the main tourism motivations (Li, Lai et al., 2011). Shopping expenses accounted for the highest proportion of China's outbound travel expenses (China Tourism Academy, 2018). Eighty-nine percent of Chinese tourists participated in shopping when travelling to the United States in 2015 (Meng et al., 2019). Although people recognize the universality and importance of tourism shopping, this aspect has not yet been fully explored (Jin et al., 2019). Existing research on outbound tourism shopping is mainly conducted at the micro level, and studying China's outbound tourism shopping behaviour at a macro level is conducive to policy-making and destination management decisions. What are the shopping behaviours of Chinese tourists visiting different countries? Why is there such a difference in behaviour? This chapter provides some preliminary insights on these questions. This chapter begins with the current research status of China's outbound tourism shopping behaviour and then introduces the content analysis research method. Through the analysis of online tourism shopping reviews, the research results are obtained and discussed.

DOI: 10.4324/9781003121329-17

Literature review

The research on China's outbound tourism mainly focuses on the pre-departure, outbound tourism destination selection (Li, Zhang et al., 2011), travel and destination perceptions (Pearce et al., 2013) and post-travel, destination satisfaction (Liu et al., 2015), involving socio-politics (King & Tang, 2009), economics (Lim et al., 2020), psychology (Chow & Murphy, 2011; Li, Zhang et al., 2011), marketing (Li et al., 2015), and management (King et al., 2006). Most studies adopt empirical or post-empirical paradigms, reflecting the pursuit of objectivity (Jin & Wang, 2015; Law et al., 2016).

Scholars have long studied the selection of outbound tourist destinations. Most scholars believe that social policy is a key element that influences the choice of China's outbound tourism destinations. Guo et al. (2007) analysed the factors affecting the Chinese Mainland's outbound tourism market from the aspects of policy, economy, and social politics by collecting tourism statistics, tourism literature, and in-depth interviews. Qu and Lam (1997) pointed out that "disposable income per capita" and "relaxation of visa requirements" will affect mainland tourists to travel to Hong Kong.

Lim and Wang (2008) analysed the arrival patterns of Chinese tourists to Australia from 1984 to 2004 and concluded that government policies could affect the number of outbound tourists. The research on destination perceptions of outbound tourism is relatively recent. Compared with short-distance trips, potential long-distance travellers care more about the function and quality of the product (Lo & Lam, 2004). The main considerations for Chinese tourists are food and accommodation. They attach great importance to cleanliness, safety and value, and long for real respect and hospitality (Li, Lai et al., 2011). In terms of destination satisfaction, Yang et al. (2016) found that word of mouth will affect tourist satisfaction.

Outbound tourism shopping behaviour

To date, researchers have mainly studied China's outbound tourism shopping at the micro level. The data collection procedures include the questionnaire survey method (Lin, 2016), interviews, and online travel community posts and mainly focus on the motivations for outbound tourism shopping behaviour and purchasing products. Meng et al. (2019) investigated the factors that influence impulsive tourism shopping behaviour from three aspects: the intrinsic attributes of tourism, social influence, and product attributes. Mainland Chinese tourists like electronic and brand-name products (Guo et al., 2007).

Tourist luxury shopping behaviour is social, and sociality not only depends on what they buy but to a large extent depends on where they shop (Correia et al., 2017). Guo et al. (2009) found that the motives of outbound tourists in Shanghai for shopping include function, emotion, and gift giving. In addition to conventional research, more scholars have recently combined

China's unique culture to study the shopping behaviour. Wang et al. (2020) used an experimental method to investigate the influence of Chinese culture on outbound tourism shopping and found that the 88% discount could stimulate Chinese tourists to buy gifts. Gao et al. (2017) found that face will affect China's outbound group buying.

For qualitative analysis, various research methods are used with outbound tourism shopping behaviour, including content analysis, interviews, and focus group discussions. Jin et al. (2019) explored the main characteristics of Chinese tourist shopping in Australia. Xu and McGehee (2012) used interviews and found that gifts for relatives and friends, brand, quality, price, and leisure time were the main motivations for shopping by Chinese tourists. Li and Ryan (2018), through in-depth interviewing, confirmed that giving gifts to family and friends was still a key driving force for Chinese tourists. Hung et al. (2018) using panel discussions found that the consumption motivation of luxury goods was oriented to society and individuals. Hung et al. (2021) studied the preferences, attributes, and potential dimensions of Chinese luxury consumers for shopping destinations. Zhu et al. (2016) used travel gaze to explain the deeper mentality behind Chinese consumer purchases and determined that most of the Mainland Chinese tourists liked to buy functional and decorative goods and social gifts when travelling in Europe.

In summary, most of the research on Chinese outbound tourism shopping behaviour has a focus on destination selection, and this shopping needs to be studied in a wider range of destinations. Therefore, this chapter explores differences in tourism shopping behaviour in various outbound countries and the reasons for dissimilar buying patterns. The Chinese outbound tourism shopping market is thought to be different from other markets (Li, Lai et al., 2011; Zhu et al., 2016). Tourism shopping behaviour is exogenous, mainly due to inner psychological levels. The research on the social psychology of Chinese tourists should be combined with local and environmental knowledge, with local thinking, to propose a new theoretical framework (Dissanayake, 2003). Shopping activity review data have the characteristics of authenticity and local thinking, so this chapter selected the online comments of Chinese outbound tourists for analysis.

Involvement theory

Involvement refers to the correlation between the product and an individual's intrinsic needs, interests, and values from a personal perspective (Shim & Kotsiopulos, 1992). Involvement theory has been applied to social psychology, marketing (Briliana et al., 2020), and advertising (Cheng et al., 2018). Researchers have put forward various consumer behaviour models and participation concepts (Huang et al., 2010). Involvement can be divided into situation, enduring, and response involvement (Houston, 1978). Situation involvement refers to consumers' attention to objects in a specific situation (Huang et al., 2010). Enduring involvement is constant attention to things at the personal level. Response involvement is a mental state that combines situation and enduring involvement.

Bergadaà et al. (1995) defined involvement in shopping as a lasting motivation to participate and suggested that the degree of a person's participation in shopping is more psychologically than socially influenced. Shim and Kotsiopulos (1992) found that travellers with high shopping participation tended to pay attention to brands, trends, and like to buy in specialty stores. Consumer participation is a function of the risk the consumers feel during the purchase decision process (Kim, 2018). Increased shopping participation directly increases cognitive impulse buying (Sohn & Lee, 2016). Kinley et al. (2010) discovered that the involvement of clothing shopping is related to how often consumers ask others for opinions. The level of consumer participation varies with different personality traits (Lin et al., 2019). Scholars hold differing views on the situation and enduring involvement tendencies of shopping participation. This chapter introduces the three tendencies of Houston's (1978) involvement theory into China's outbound tourism shopping behaviour to examine the variations in and reasons for shopping behaviour in different countries.

Methodology

This research employed content analysis to determine the differences in shopping behaviours and the reasons for these dissimilarities in the top ten destination countries for Chinese outbound tourism. The ten most popular countries for Chinese overseas shopping in 2019 were Japan, UAE, UK, France, Singapore, USA, Spain, South Korea, Italy, and Australia (Ctrip & International, 2019). Ctrip[1] is China's largest and most popular online travel platform, containing a large number of tourist reviews. Therefore, Ctrip was selected as the research source to collect the review text information on China's outbound tourism shopping behaviour.

The method was divided into two stages: data collection and analysis. In the data collection, the Octopus collector was used to crawl the review information of shopping in the ten countries, including review text, shopping scoring, and review time. For the data analysis, purposeful sampling was adopted, focusing on generating enough codes to achieve theoretical saturation (Suri, 2011). In the first stage of data analysis, 1,000 review texts from each country were selected for one by one reading. The selection criteria were also factors such as shopping attraction types, text information, and review time. The textual data were encoded and categorized during reading. Table 14.1 shows some examples of data encoding. In the second stage, it was checked whether more comment texts could provide new insights into shopping behaviour. At this stage, the researchers repeated the same data analysis process as in the first stage. During the coding and classification process, the researchers repeatedly saw similar examples without any new categories, and there was no need to collect and analyse further data.

Table 14.1 Data extracts with codes, categories, and themes.

Data extracts	*Codes*	*Categories*	*Themes*
The department store is good at Takashimaya, and the price is still expensive. Getting off at Nihonbashi Subway Station is the closest.	Takashimaya Expensive Subway	Native style (1) Price (2) Transport (3)	Situation factors (1) Product factors (2, 3)
Every time I go to a new city, I like to buy a few special postcards. This habit is greatly satisfied here.	Like Satisfy	Interest (1) Emotion (2)	Personal factors (1, 2)

Results

This section is divided into two parts to introduce the differences in shopping behaviours of Chinese tourists in the top ten outbound countries and the reasons for these dissimilarities. The first part explains the main characteristics and differences of Chinese tourist shopping behaviour in the top ten outbound countries. The second part focuses on analysing the reasons for varying shopping behaviours in different countries.

Main characteristics and differences

Shopping elements: Shopping elements refer to the content that tourists pay attention to when shopping, which can be divided into services, transportation, brands, characteristics, price, quality, and emotions. The elements of shopping in different countries are quite different in these dimensions.

In terms of services, Chinese shopping guides in South Korea have brought great convenience to shoppers. The South Korean Attractions Review wrote, "Every store must have a Chinese shopping guide to follow and provide caring services." There are many service centres in Spain. For example, "the most important thing is service. The service center in La Roca is bigger and more beautiful, with good service attitudes." The same goes for transportation. In Japan, it is very convenient to take the subway between shopping attractions. For example, "It is located at the Shinjuku Transportation Hub Station, and many subway lines such as JR Line and Toei Line are directly accessible, very convenient!" Taking a bus in Spain is considered "green."

Almost all Chinese tourists pay attention to the brand of the product, but there are variations in different countries. In Japan, local Japanese brands that are close to the people are more likely to be bought. As one tourist said, "Those luxury brands are built from building to building. Buying Japanese local brands is the most cost-effective way, such as Uniqlo and Muji. Japan has some styles that are not available in China." Just like "a street full of luxury goods, buy, buy, buy," tourists in France are more inclined to buy luxury brands.

For features, products with local characteristics are easier to buy. For example, "This is a must-visit in England. It is recommended that you not only visit the big luxury brands, but also the qualities of many British brands are worth discovering." In terms of prices, most tourists seek lower prices when shopping. Among them, tax exemptions are a priority concern. For example, "Bring your passport and get tax exemptions directly." Second, credit card discounts and gifts are also factors that stimulate tourists to purchase. In terms of quality, the quality of shopping in almost all outbound countries is recognized. For example, "These are all world-class brands, and the quality of the products can be assured." Since outbound tourists mainly shop for high-luxury products, the quality is guaranteed. However, there are also people who have doubts about the quality. For example, in Dubai, "the quality of gold is uneven, so be cautious when you start." In terms of emotions, shopping in most countries is mainly for entertainment. In addition, a large part of buying luxury goods in France is to show off (Table 14.2). For example, "Lancel's window on the Champs Elysée is more artistic than LV's naked show of luxury." Many people shopping in the United States do so out of social needs, such as "friends and moms' cosmetics are purchased here."

Shopping categories: Shopping categories can be divided into functional and decorative goods and social gifts. In Japan, USA, and Australia, tourists preferred to purchase functional products such as Japanese electrical appliances, American refrigerators, and Australian lanolin. In almost all Western outbound countries, tourists will purchase decorative goods. In Europe, they preferred to buy clothing and cosmetics. In the UAE, they were fond of buying gold and spices. For example, "the interior is brilliant, gold has many styles, and the price is relatively cheap, not bad." In France, there are more purchases of bags, and in Australia, people tended to buy woollen accessories. People mainly bought social gifts in Europe, America, and Oceania, while buying less in neighbouring countries in Asia. The special attributes of gifts were considered when purchasing (Table 14.3).

Shopping situations: Shopping situation refers to tourist choices and perceptions of the environment during the shopping process. For example, "If you have time and love to go shopping, you must go shopping on Oxford Street. It is not for the purchase, but the shopping atmosphere is really good." Shopping context can be divided into two categories: native style and commercial atmosphere. Almost all outbound countries have a strong commercial atmosphere, and the local shopping styles of countries are very different. For example, local shopping in Japan and South Korea is more inclined to create a lively atmosphere, which helps to stimulate the herd mentality, and makes tourists more willing to go shopping where locals shop. For example, "There are still more locals, you can eat with confidence." In Britain and France, the Chinese prefer the style of famous stores, which offer famous branded products such as Harrods (Table 14.4).

Table 14.2 Differences in shopping elements in different countries

Country	*Services*	*Transportation*	*Brands*	*Characteristics*	*Price*	*Quality*	*Emotions*
Japan	Chinese shopping guide	Subway	Local brands	Close to the people	Low price	Good	Relaxed
UAE	Good attitudes	Bus, boat, parking lot	Luxury brands	Luxurious	Low price	Uneven	Happy
UK	English communication	Subway, bus	Luxury brands, British brands	British flavour	Low price	Good	Happy
France	Chinese shopping guide	Subway	Luxury brand, European, and American high street brands	Fashion	High prices	Good	Show off
Singapore	Chinese shopping guide	Subway	Luxury brands, Popular brands	Designed	Low price	Good	Happy
USA	Chinese shopping guide	Subway	Luxury brands	Buildings	Average	Good	Social
Spain	Service centres, guide	Bus	Local brand, first-tier big names	Art	Low price	Good	Casual
South Korea	Chinese shopping guide	Subway	Local brands, trendy brands	Lively	Low price	Uneven	Happy
Italy	Chinese shopping guide	Bus	Luxury brands	Fashion	Low price	Uneven	Happy
Australia	Chinese shopping guide	Train, tram	Local brands luxury brands	Culture	Low price	Good	Culture

Table 14.3 Differences in shopping categories in different countries

Country	*Functional goods*	*Decorative goods*	*Social gifts*
Japan	Electrical appliances, seafood	Clothing, cosmetics	Less
UAE	Less	Accessories, gold, spices	Less
UK	Less	Clothing	Souvenirs
France	Macaroon	Clothing, cosmetics, perfumes, bags	Souvenirs
Singapore	Food	Clothing	Souvenir
USA	Refrigerators, chocolate	Cosmetics	Souvenirs, specialty
Spain	Ham, seafood	Crafts	Souvenirs
South Korea	Snacks, seafood	Clothing, cosmetics	Less
Italy	Seafood	Clothing, perfumes, masks	Souvenirs
Australia	Lanolin, seafood	Wool products	Souvenirs

Table 14.4 Differences in shopping situations in different countries

Country	*Native or local style*	*Commercial atmosphere*
Japan	Animation shop, lively places	Shopping malls, duty-free shops
UAE	Old town	Shopping malls
UK	British stores	Shopping malls, shopping villages, shopping streets
France	Famous stores, fashionable, and romantic places	Specialty stores
Singapore	Chinatown, specialty shops	Shopping malls, duty-free shops
USA	Supermarkets, markets	Duty-free shops
Spain	Markets	Walking streets
South Korea	Markets, small shops, lively places	Commercial streets, walking street, duty-free shops
Italy	Markets	Commercial streets, flagship stores, department stores
Australia	Old houses, Queen's buildings	Shopping malls

Reasons for differences

Through the content analysis method to code, this study found that the outbound travel shoppers had different shopping behaviours due to personal, product, and situational factors. Personal factors are internal, product are external, and situational factors are immediate or environmental variables.

Personal factors: The degree of consumer participation varies with different personal factors (Lin et al., 2019). Personal factors include emotions, hobbies, and demands. Due to individual differences in these three aspects, shopping involvement varies, and shopping behaviours are dissimilar. Different emotions lead to variations in shopping behaviours. For example, out of the psychology of showing off, it is more likely to form impulsive consumption at a low level of involvement and then purchase luxury goods. Due to social psychology, people spend more time in selecting products and are more involved, so they are more likely to buy specialty products. In terms of hobbies, for example, "I am not particularly interested in Japanese clothes. Instead, I like some of their daily necessities like towels, handkerchiefs, and socks." This reviewer was more involved in daily necessities due to his personal hobbies and was easier to buy such products. In terms of demand, it is mainly due to differences in personal needs that lead to differences in shopping. For example, "When I was in Paris, the masks just ran out, so I went and bought some for emergency." "There are so many delicious and fun things that can definitely meet the needs of buying and buying."

Product factors: Product factors refer to the services, transportation, brands, characteristics, prices, and quality of products. Because different countries have very dissimilar products, the shopping behaviour of tourists is also variable. Taking transport as an example, choosing different modes when shopping determines the quantity and type of purchase to a certain extent. For example, people who take public transportation such as subways and buses for shopping prefer small-volume products. For brands, the development of Japanese local brands is very good, so when tourists shop in Japan, they will buy more local brands, especially local electronic appliances and cosmetics. In France, Italy, and the UK, upscale and famous brands are well-known and highly appealing, so tourists are more likely to spend on luxury brands. In terms of prices, low-price behaviour will stimulate tourists to purchase, although this does not apply in all countries. For example, in the UAE, low-priced gold causes people to doubt the quality, and low prices in France will not support a show-off psychology.

Situational factors: Situational factors are those that have an impact on an individual's behaviour or psychology at a certain time or in a particular environment and are divided into commercial atmosphere and local style. Having a commercial atmosphere is an essential element for shopping abroad, and local styles produce different shopping behaviours. As one commenter said, "An indelible shopping truth is to buy local products locally. Buying British products in the UK must be cheaper than elsewhere." Shopping behaviour in the destination country is related to the country's unique styles. For example, Chinese people tend to go to department stores when shopping in Japan and buy Japanese medicine and cosmetics. Commercial atmosphere is an essential element for shopping. Tourists may buy more products in places with strong commercial atmosphere. For example, "Myeongdong is a shopping mall in Seoul, very prosperous and stylish. It is a good place to buy in Seoul."

Conclusion

This research applied involvement theory in the study of outbound tourism shopping behaviour and explored the differences in and reasons for Chinese tourist shopping in the top ten outbound countries. It was found that the combined influence of personal, product, and situation factors through situation involvement, enduring involvement, or response involvement (Houston, 1978) led to dissimilarities in shopping elements, categories, and situations in different countries (Figure 14.1).

Situation involvement refers to consumers' attention to objects in a specific situation. Tourists feel the extravagant atmosphere in France and form irrational consumption through situational involvement, which ultimately leads to the purchase of luxury goods. For example, "Unlike the large department stores all over the world, the Galeries Lafayette department store in Nice has also started a Christmas shopping spree. The quality of goods, service quality, and prices here will tempt you to buy and buy." This confirms the view of Correia et al. (2017), who said that "tourists' luxury shopping behaviour is social, and this not only depends on what they buy, but to a large extent, it depends on where they shop."

Enduring involvement is constant attention to things at the personal level. Buying daily necessities in Japan is an example of enduring involvement. As a result of personal hobbies or demands, people prepare strategies for purchasing products in advance and draw up a shopping list. For example, "I did a lot of homework before going to Japan. I also made a long shopping list."

Response involvement is a mental state that combines situation and enduring involvement. Hung et al. (2018) found that the consumption motives of luxury goods are both social and personal. This chapter expands this notion from luxury goods to outbound shopping products and suggests that shopping can be promoted by response involvement. People will not only

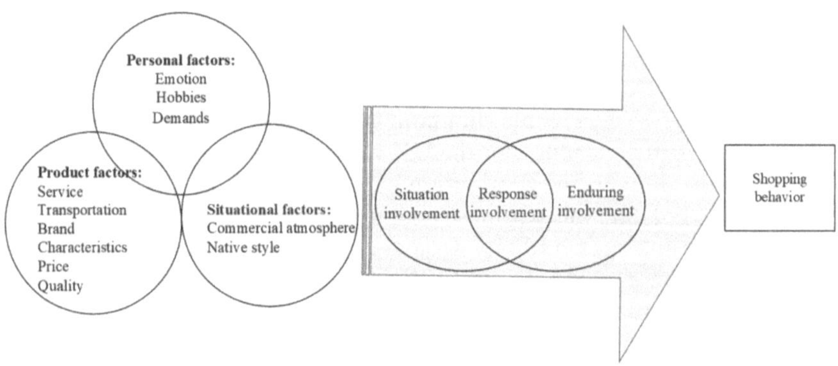

Figure 14.1 Tourism shopping involvement model

prepare shopping strategies in advance but are also be affected by the specific situation at the time and exhibit shopping behaviours under the combined effect of situation and enduring involvement. For example, "London clothes are of average quality, they are all made in China, the price is not cheap, and the materials are not good… But if you have time, if you like to go shopping, you must go to the Oxford Street, not for how much to buy, but the atmosphere of shopping is really good." It can be seen that the tourists had already prepared a strategy before buying clothes in London and experienced the atmosphere at that time while shopping. The combined effect of product and situational factors will ultimately produce the purchase behaviour.

The results have some theoretical contributions. First, this research examines the shopping behaviour of the top ten popular outbound countries for Chinese tourists. This is in sharp contrast with the previous research and is of great significance for understanding China's outbound tourism market. Second, this research has enriched the application of involvement theory in the tourism shopping context and proposed a tourism shopping involvement model, which is appropriate for understanding the differences and causes of tourism shopping in various countries.

The chapter also has certain implications for tourism marketing and management. First, with the continuous expansion of China's outbound tourism market and shopping consumption, the outbound shopping of Chinese tourists deserves more attention from destination marketing organisations. The results will help DMOs to better understand Chinese outbound tourists so as to provide targeted products and services to improve the shopping experiences of these tourists. Second, the tourism shopping involvement model suggests that varying personal, product, and situational factors leads to different shopping behaviours. Marketers can stimulate shopping through various aspects such as emotion creation, product attributes, and situational experiences. For example, creating a situation with local characteristics is more conducive to promoting the sales of local products, while creating a luxurious and gorgeous atmosphere promotes impulsive consumption and increases the purchase of luxury goods.

This research had certain limitations. First, the data were from online review texts on Ctrip, and these comments were mainly written by tourists who were travelling independently. Group shopping should be explored in the future. In addition, this research provides some preliminary indications of the outbound tourism shopping phenomenon, but the exploration of the relationships of various factors requires more in-depth consideration. In the future, quantitative research methods should be used to verify the model and more robustly explore the relationships within the model.

Note

1 www.ctrip.com.

References

Bergadaà, M., Faure, C., & Perrien, J. (1995). Enduring involvement with shopping. *The Journal of Social Psychology, 135*(1), 17–25. doi:10.1080/00224545.1995.9711396

Briliana, V., Deitiana, T., & Mursito, N. (2020). Exploring antecedents of social media involvement in shopping online for Muslim fashions: Evidence from Indonesia. *PalArch's Journal of Archaeology of Egypt/Egyptology, 17*(7), 2789–2800.

Cheng, Z.-H., Chang, C.-T., & Lee, Y.-K. (2018). Linking hedonic and utilitarian shopping values to consumer skepticism and green consumption: The roles of environmental involvement and locus of control. *Review of Managerial Science, 14*, 61–85. doi:10.1007/s11846-018-0286-z

China Tourism Academy. (2017). *Annual report of China outbound tourism development 2017.* Beijing: Tourism Education Publishing.

China Tourism Academy. (2018). *Annual report of china outbound tourism development 2018.* Beijing: Tourism Education Press.

China Tourism Academy. (2020). *China outbound tourism development report (2020).* http://www.199it.com/archives/1151221.html

Chow, I., & Murphy, P. (2011). Predicting intended and actual travel behaviors: An examination of Chinese outbound tourists to Australia. *Journal of Travel & Tourism Marketing, 28*(3), 318–330. doi:10.1080/10548408.2011.563166

Cohen, E., & Cohen, S. A. (2012). Current sociological theories and issues in tourism. *Annals of Tourism Research, 39*(4), 2177–2202. doi:10.1016/j.annals.2012.07.009

Correia, A., Kozak, M., & Kim, S. (Sam). (2017). Luxury shopping orientations of Mainland Chinese tourists in Hong Kong. *Tourism Economics, 24*(1), 92–108. doi:10.1177/1354816617725453

Ctrip & International. (2019). *New tourism, new consumption, new middle class: A report on Chinese outbound Tourism consumption in 2019.* http://www.lvjie.com.cn/brand/2019/1129/15028.html

Dissanayake, W. (2003). Asian approaches to human communication: Retrospect and prospect. *Intercultural Communication Studies, 12*(4), 17–38.

Gao, H., Huang, S. (Sam), & Brown, G. (2017). The influence of face on Chinese tourists' gift purchase behaviour: The moderating role of the gift giver–receiver relationship. *Tourism Management, 62*, 97–106. doi:10.1016/j.tourman.2017.03.013

Guo, Y., Pei, Y., Ye, Y., Chen, Y., Wang, K.-C., & Chan, H.-C. (2009). Tourist shopping behavior: A case of Shanghai outbound tourists. *Journal of Tourism, Hospitality & Culinary Arts (JTHCA), 1*(1), 1–18. https://core.ac.uk/download/pdf/154828437.pdf

Guo, Y., Seongseop Kim, S., & Timothy, D. J. (2007). Development characteristics and implications of Mainland Chinese outbound tourism. *Asia Pacific Journal of Tourism Research, 12*(4), 313–332. doi:10.1080/10941660701760995

Houston, M. J. (1978). Conceptual and methodological perspectives on involvement. In S. Jain (Ed.), *Research frontiers in marketing: Dialogues and directions* (pp. 184–187). Chicago, IL: American Marketing Association.

Huang, C.-Y., Chou, C.-J., & Lin, P.-C. (2010). Involvement theory in constructing bloggers' intention to purchase travel products. *Tourism Management, 31*(4), 513–526. doi:10.1016/j.tourman.2009.06.003

Hung, K., Qiu Zhang, H., Guillet, B. D., & Wang, L. (2018). China watching: Luxury consumption and its implications. *Journal of Travel & Tourism Marketing*, 577–592. doi:10.1080/10548408.2018.1525470

Hung, K., Ren, L., & Qiu, H. (2021). Luxury shopping abroad: What do Chinese tourists look for?. *Tourism Management*, *82*, 104182. doi:10.1016/j.tourman.2020.104182

Jin, H., Moscardo, G., & Murphy, L. (2019). Exploring Chinese outbound tourist shopping: A social practice framework. *Journal of Travel Research*, 1–17. doi:10.1177/0047287519826303

Jin, X., & Wang, Y. (2015). Chinese outbound tourism research: A review. *Journal of Travel Research*, *55*(4), 440–453. doi:10.1177/0047287515608504

Kim, J.-H. (2018). Effect of brand popularity as an advertising cue on tourists' shopping behavior. *Journal of Destination Marketing & Management*, *10*, 78–86. doi:10.1016/j.jdmm.2018.07.001

King, B., Dwyer, L., & Prideaux, B. (2006). An evaluation of unethical business practices in Australia's China inbound tourism market. *International Journal of Tourism Research*, *8*(2), 127–142. doi:10.1002/jtr.563

King, B., & Tang, C. H. (2009). China's outbound tourism during the 1980s—A socio-political perspective. *Anatolia*, *20*(1), 18–32. doi:10.1080/13032917.2009.10518892

Kinley, T. R., Josiam, B. M., & Lockett, F. (2010). Shopping behavior and the involvement construct. *Journal of Fashion Marketing and Management: An International Journal*, *14*(4), 562–575. doi:10.1108/13612021011081742

Law, R., Sun, S., Fong, D. K. C., Fong, L. H. N., & Fu, H. (2016). A systematic review of China's outbound tourism research. *International Journal of Contemporary Hospitality Management*, *28*(12), 2654–2674.

Li, F. (Sam), & Ryan, C. (2018). Souvenir shopping experiences: A case study of Chinese tourists in North Korea. *Tourism Management*, *64*, 142–153. doi:10.1016/j.tourman.2017.08.006

Li, M., Zhang, H., Mao, I., & Deng, C. (2011). Segmenting Chinese outbound tourists by perceived constraints. *Journal of Travel & Tourism Marketing*, *28*(6), 629–643. doi:10.1080/10548408.2011.598747

Li, X. (Robert), Cheng, C.-K., Kim, H., & Li, X. (2015). Positioning USA in the Chinese outbound travel market. *Journal of Hospitality & Tourism Research*, *39*(1), 75–104. doi:10.1177/1096348012461548

Li, X. (Robert), Lai, C., Harrill, R., Kline, S., & Wang, L. (2011). When east meets west: An exploratory study on Chinese outbound tourists' travel expectations. *Tourism Management*, *32*(4), 741–749. doi:10.1016/j.tourman.2010.06.009

Lim, C., & Wang, Y. (2008). China's post-1978 experience in outbound tourism. *Mathematics and Computers in Simulation*, *78*(2–3), 450–458. doi:10.1016/j.matcom.2008.01.033

Lim, D. J., Ferguson, V. A., & Bishop, R. (2020). Chinese outbound tourism as an instrument of economic statecraft. *Journal of Contemporary China*, *29*(126), 916–933. doi:10.1080/10670564.2020.1744390

Lin, H.-W., Jhan, Y.-C., & Chao, W. (2019). Online shopping behavior of emerging shopping websites. *Saudi Journal of Business and Management Studies*, *4*(3), 207–214.

Lin, L. (2016). Food souvenirs as gifts: Tourist perspectives and their motivational basis in Chinese culture. *Journal of Tourism and Cultural Change*, *15*(5), 439–454. doi:10.1080/14766825.2016.1170841

Liu, X., Li, J. J., & Yang, Y. (2015). Travel arrangement as a moderator in image–satisfaction–behavior relations: An investigation of Chinese outbound travelers. *Journal of Vacation Marketing*, *21*(3), 225–236. doi:10.1177/1356766714567797

Lo, A., & Lam, T. (2004). Long-haul and short-haul outbound all-inclusive package tours. *Asia Pacific Journal of Tourism Research, 9*(2), 161–176. doi:10.1080/1094166042000233694

Meng, F., Zhang, P., Li, H., & So, K. K. F. (2019). Modeling precursors of impulsive tourist shopping behavior: Evidence from long-haul Chinese outbound tourists. *International Journal of Tourism Research*, 1–15. doi:10.1002/jtr.2266

Pearce, P. L., Wu, M.-Y., De Carlo, M., & Rossi, A. (2013). Contemporary experiences of Chinese tourists in Italy: An on-site analysis in Milan. *Tourism Management Perspectives, 7*, 34–37. doi:10.1016/j.tmp.2013.04.001

Qu, H., & Lam, S. (1997). A travel demand model for Mainland Chinese tourists to Hong Kong. *Tourism Management, 18*(8), 593–597. doi:10.1016/s0261-5177(97)00084-8

Shim, S., & Kotsiopulos, A. (1992). Patronage behavior of apparel shopping: Part I. Shopping orientations, store attributes, information sources, and personal characteristics. *Clothing and Textiles Research Journal, 10*(2), 48–57. doi:10.1177/0887302x9201000208

Sohn, H.-K., & Lee, T. J. (2016). Tourists' impulse buying behavior at duty-free shops: The moderating effects of time pressure and shopping involvement. *Journal of Travel & Tourism Marketing, 34*(3), 341–356. doi:10.1080/10548408.2016.1170650

Suri, H. (2011). Purposeful sampling in qualitative research synthesis. *Qualitative Research Journal, 11*(2), 63–75. doi:10.3316/qrj1102063

Wang, D., Martin, B. A. S., & Yao, J. (2020). Do discount presentations influence gift purchase intentions and attitudes of Chinese outbound tourists?. *Journal of Travel Research, 60*, 1–19. doi:10.1177/0047287520918008

Wang, Y., & Davidson, M. C. G. (2010). Chinese holiday makers' expenditure: Implications for marketing and management. *Journal of Hospitality Marketing & Management, 19*(4), 373–396. doi:10.1080/19368621003667101

Xu, H., Wang, K., & Song, Y. M. (2018). Chinese outbound tourism and soft power. *Journal of Policy Research in Tourism, Leisure and Events, 12*, 34–49. doi:10.1080/19407963.2018.1505105

Xu, Y., & McGehee, N. G. (2012). Shopping behavior of Chinese tourists visiting the United States: Letting the shoppers do the talking. *Tourism Management, 33*(2), 427–430. doi:10.1016/j.tourman.2011.05.003

Yang, Y., Jing, F., & Nguyen, B. (2016). China's outbound tourism: Investigating word-of-mouth and its effect on perceived value, satisfaction and loyalty. *Journal of China Tourism Research, 12*(2), 159–178. doi:10.1080/19388160.2016.1223776

Zhu, D., Xu, H., & Jiang, L. (2016). Behind buying: The Chinese gaze on European commodities. *Asia Pacific Journal of Tourism Research, 21*(3), 293–311. doi:10.1080/10941665.2015.1048263

15 Who are Chinese luxury travelers?

Elaine Yulan Zhang and Tony S. M. Tse

Introduction

Chinese luxury travel market is attractive for its market size. The number of high-net-worth individuals (HNWI) in China reached 1.97 million in 2018 (Statista, 2019), and 3% of Chinese travelers are willing to spend 50,000 RMB (US$7,316) or more per trip (Ctrip Tourism Big Data Lab, 2020, as cited in Globetrender, 2020). "Chinese luxury travelers" are often discussed among industry players, for example, "what do Chinese luxury travelers want" (Jing Culture & Commerce, 2020) and "the rise of Chinese luxury travelers" (International Luxury Travel Market [ILTM], 2019). The meaning of "Chinese luxury travelers," however, is not clear. When we talk about Chinese luxury travelers, who are we talking about? Having "Chinese luxury traveler" in a discussion may not suggest the same subject matter. Is financial ability the only common attribute in defining luxury travelers? Are there other pertinent characteristics? How do luxury travelers perceive themselves? This chapter explores the underlying meanings of "Chinese luxury travelers" from the lenses of researchers, practitioners, and travelers themselves.

Literature review

Chinese luxury travelers have been studied in industrial reports with definitions or screening criteria regarding wealth level, travel expenditure, and the type of service or products they consume. Jing Culture & Commerce (2020) have used "Chinese luxury travelers" and "Chinese affluent travelers" interchangeably. Hurun Report (2018) considers travelers with high travel expenditure as luxury traveler and used annual travel expenditure (350,000 RMB/US$53,330) as screening criterion. ILTM (2019) considers them as clients of luxury travel agents, travelers with high spending and often combining travel with other investments. Due to lack of agreed-upon definition of luxury travelers (Iloranta, 2019), there is a concern that "Chinese luxury traveler" becomes a label without common and essential meaning, as pointed out in the Ritz-Carlton Luxury Meeting that the word "luxury" is

DOI: 10.4324/9781003121329-18

an abused word used in the hospitality industry (MeetingsNet, 2006). This concern warrants further research on the meanings of "Chinese luxury travelers."

Academic research on Chinese luxury travelers is limited in number probably due to the difficulty in accessing luxury travelers (Zhang & Tse, 2018). These studies either used the types of service or products such as hotel or tour as screening criteria or carried out the research without a clear definition. Wu and Yang (2018) recruited travelers who have stayed in luxury hotels and investigated the values associated with their purchase intention such as hedonic value. Huang et al. (2020) conducted survey among Chinese travelers who had cruise experience and found quality perception as an important contributor of luxury value. Chen et al. (2019) surveyed among high-end travelers who joined high budgeted tours (e.g., 120,000 RMB/US18,400 for a tour to Europe) and found that these travelers prefer direct flight, tour with less designated shops, and are not price-sensitive. Based on literature related to luxury travel, travel motivation, and luxury consumption, Zhang and Tse (2018) identified 17 potential motivations of Chinese luxury travelers such as status without defining the segment.

The above literature on Chinese luxury travelers from industry perspective focused more on "what"-preferences and behaviors such as popular destinations (e.g., Hurun Report, 2018). The academic researchers were more interested in "why"-motivations of consuming luxury travel products (e.g., Wu & Yang, 2018). Similar pattern can be observed among luxury-traveler-related studies conducted in other contexts. For example, Virtuoso (2019), a global network for luxury travel agents, conducted survey among travel advisors and predicted the global trend of for luxury travelers' preference, while an academic research published in the same year looked into underlying motivations of luxury consumption among travelers such as uniquess and status (Correia et al., 2019). The various definitions of luxury travelers range from quantified definitions such as top 1% of household income owners (Resonance Consultancy Ltd., 2016) to soft definitions such as travelers who choose luxury travel as a norm (Bakker, 2005). The existing literature seems to reflect the different characteristics of the same group of people, and it is worth exploring the common attributes of "Chinese luxury travelers."

Methodology

This study uses data triangulation as it encourages tidying data into themes to increase credibility and validity of research (Savin-Baden & Major, 2013). Three sets of data include: (1) reviewing academic literature and industrial reports which represent researchers' perspective, (2) interviewing luxury travel agents who represent practitioners' perspective, and (3) interviewing luxury travelers. The first set of data includes academic papers and industrial reports which have been collected on 29 November 2020. There were 28, 151, and 903 results from Google Scholar using key words "Chinese luxury travelers," "luxury traveler," and "luxury tourist," respectively, and 143 results from Google using key words "luxury traveler" and "report." Articles focusing solely on luxury shopping during travel without discussion of the concept "luxury travelers" were not collected. Since the articles are ranked by relevance by Google and there were no relevant articles identified after the first 30 results, the manual screening process was stopped at the 50th result. A total of 22 related articles have been collected, including three book chapters, eight journal articles, one patent document, and ten reports.

The other two sets of data include interviews with 11 Chinese luxury travel agents and 30 Chinese travelers. The interviews with agents were conducted during December 2016 to March 2017. The companies they work for are listed as Luxury Travel Agents by Hurun Report or Conde Nast Traveler. Luxury travel agents were asked to share their understanding of luxury travel and luxury travelers, for example, "what makes luxury travel products differ from ordinary travel product?" and "could you categorize your clients?" The third set of data includes interviews with travelers whose annual travel expenditure was not less than RMB100,000 (US$ 14,500) and who took two business/first-class flights in the past three trips. The participants were recruited by either research company or researchers' personal connections during March 2017 to January 2018. The travelers were explained that they were participating a research among Chinese luxury travelers and then asked to share their past travel experience and dream trip in future, for example, "describe the most worth sharing or most recent trip?" and "may I know your dream trip?" All the collected articles and interview transcripts are imported to NVivo for content analysis.

Results

In academic works and industrial reports, luxury travelers are defined by demographics (e.g., wealth), behaviors (e.g., traveling with helicopter), characteristics of experience (e.g., exclusive), and values travelers look for (e.g., status). Majority of academic works (9 out of 11) and industrial reports (9 out of 11) are searching for the common characteristics (Table 15.1). For example, the report by Resonance Consultancy Ltd. (2016) revealed that luxury travelers are alike, and their desire for "exclusivity, customization, and options" has "nothing to do with the environment or nationality." While more travelers start to consume luxury travel products, papers or reports acknowledge differences among travelers. For example, Amadeus (2016) identified six types of luxury travelers: always luxury who consider luxury as the minimum requirement; special occasion who sacrifice luxury facilities for special experience; luxury who extend business trip for luxury leisure experience; cash-rich, time-poor who often make last-minute decision and use agent; strictly opulent who share their experience on social media; and independent and affluent who would like to pamper themselves and try something new.

Table 15.1 Summary of academic papers and industrial reports

	Academic works n = 11	*Industrial reports n = 11*
Focus: common characteristics	9	9
Focus: segments within luxury travelers	2	2
Demographics: wealth/income	3	7
Demographics: elites/CEOs	0	4
Behaviors: products (e.g., travel advisor and top hotel rooms)	4	10
Behaviors: activities (e.g., fitness and shopping)	4	10
Trip characteristics (e.g., authentic and exclusive)	7	11
Values: short-term (e.g., comfort and safety)	3	3
Values: long-term (e.g., self-actualization and status)	6	6

Demographically, luxury travelers are often described by their wealth and occupation, and people who are lack of financial support but have a taste of "luxury traveler" is also included (Popescu & Olteanu, 2014). However, almost studies consider strong financial ability as one important characteristic of luxury travelers. Terms like "affluent travelers" (Fairmont Hotels & Recorts, 2018) and "HNW travelers" (Horwath HTL, 2018) are used to describe luxury travelers. High income was often used as a criterion of selecting or describing participants. Annual household incomes of $200,000 or more were used in a report targeting international travelers (Resonance Consultancy Ltd., 2016). The roles of luxury travelers in daily life are described as CEOs (Amadeus, 2016), entrepreneurs, investors (HH Travel, 2016), professionals (Bakker, 2005), or a more general role as "elites" (Resonance Consultancy Ltd., 2016). Luxury travelers are defined by their wealth and occupation in their daily life. Luxury travel is an "extension of their normal lifestyle" (Swarbrooke, 2018, p. 5).

Luxury travelers can also be understood by observing their behaviors, that is, what they buy and what they do. They would need travel adviser for trip planning (Skift, 2017), book top hotel rooms (Stankova & Kaleichev, 2013), purchase or rent vacation home (Hurun Report, 2018), fly with business class, private jet, or helicopter (Amadeus, 2016), and enjoy cruise or yacht (HH Travel, 2016). CEO of Virtuoso highlighted the importance of personal travel advisor: "*luxury is having someone—an actual person—know you well enough*" (Skift, 2017). Luxury travel tends to focus on leisure activities although one study includes corporate travelers (Bakker, 2005). Luxury travelers would like to participate in wellness activities such as fitness (Resonance Consultancy Ltd., 2016), stay close to the nature such as sea view (Hurun Report, 2018), enjoy local food (Stankova & Kaleichev, 2013), go shopping (HH Travel, 2016), experience art such as art festival (Resonance

Consultancy Ltd., 2016), participate in special events such as day of dead parade in Mexico (Brenner & Aguilar, 2002), carry out extreme experiences such as climbing Kilimanjaro (Ovation Travel, 2018), visit universities (HH Travel, 2016), and join philanthropy (Ovation Travel, 2018).

As reflected by the above activities, luxury traveler's experience is exclusive (Bowerman, 2014), expensive (Bakker, 2005), authentic (Iloranta, 2019), pampered with high-quality service (Horwath HTL, 2018), and sustainability responsive (Fairmont Hotels & Recorts, 2018). The experience is also described in the following ways: not serious but approachable (Ovation Travel, 2018), often full of human contact (Horwath HTL, 2018), allowing participation (Stankova & Kaleichev, 2013), predictable in terms of service level (Ovation Travel, 2018), time saving (Bakker, 2005), and value for money for travelers who do not have sufficient financial support (Popescu & Olteanu, 2014). Specifically, Chinese traveler enjoy exclusivity which allows not "surrounded by other Chinese tourists" (Bowerman, 2014, p. 48).

The values or purposes of luxury travelers include two categories by the term of influence: short-acting and long-lasting. The short-term values include privacy (HH Travel, 2016), hedonism (Amadeus, 2016), relaxation (Novotná & Kunc, 2019), safety (Zhang & Tse, 2018), and cleanliness (Hurun Report, 2018). These values may be maximized during the trip and will disappear or be diminished after the trip. The long-term values include status (Correia et al., 2019), uniqueness (Iloranta, 2019), learning new things (Resonance Consultancy Ltd., 2016), transformation (Horwath HTL, 2018), social value (Skift, 2017), self-actualization (Iloranta, 2019), self-challenge (Ovation Travel, 2018), and individualism (Zhang & Tse, 2018). Luxury travelers should like to change the way how they look at themselves or this world through transformational experience in traveling.

Travel agents

Travel agents agree that sufficient financial support is one important characteristic of Chinese luxury travelers. These travelers may include "*nouveau riche*" (agent #2), and those who are not "*lack of money*" (agent #10). When agents were asked whether they have any guests who have insufficient financial support, they commented that "*this group of people is small*" (agent #11), and "*sometimes we cannot cater to those guests*" (agent #3). Agent #4 explained "four-star hotels" as the threshold: "*if they can only afford three-star hotels or even Airbnb, we cannot take them because we will not be able to guarantee the quality of service.*" Agents describe their guests as people who are often time poor, while their family members such as wife and kids may have free time. Some of these busy-working travelers can decide when to take the vacation: "*must be the bosses or have flexible working hours*" (agent #4).

Agents have confirmed what luxury travelers would need personal travel advice as revealed by reports and academic papers and explained reasons: limited ability or limited time. Agent #3 mentioned that travelers "*know exactly what they want. It's just a bit difficult for them to organize, and that's why they need support from experienced travel agent.*" Some travelers may have the ability to arrange the trips by themselves but they would "*rather pay for the service*" because they "*can spend small amount of money to buy time*" (agent #6). Other than travel advisor, they would also buy business classes, top hotels, private vehicle, fine-dining, luxury cruises, and helicopter. Luxury travelers buy cruise trips for their parents because there is "*no languages barrier*" and "*everything is inclusive*" (agent #2).

Chinese luxury travelers may undertake various activities ranging from polar experience to trips with kids similar to what reports and academic papers have described. One interesting concept mentioned by agent is that "*hotels are also destinations*" (agent #1). The boundary between the activities and accommodation becomes blurry since travelers may consider visiting a new hotel as the major purpose of a trip. Shopping is an important activity because "*it is smart*" to buy in overseas destination for tax reason (agent #3). The availability of the products at a destination may also be a motive for travelers to visit a place. For example, two travelers visited Japan because they wanted to purchase a Hermes handbag (agent #10).

Similar to what was commented in academic papers and industrial reports, the experience of Chinese luxury travelers is described by travel agents as exclusive, expensive, authentic, or local, served with high quality, predictable in terms of service level, and value for money. The exclusiveness of their travel experience can be demonstrated by both hardware and software. The hardware is relatively standardized: "*basically what they picked are all top hotels at the destination, all vehicles they use are Benz, all flights are business class*" (agent #9). The software depends on agents' abilities and travelers' expectation. Travelers' expectation can sometimes challenge travel agents' ability: "*regular travel products cannot meet their needs...when they request anything...we would be...wow, we have never tried*" (agent #5).

Most agents would consider the experience as expensive comparing with regular travel products. But agent #3 commented that the experience does not "*necessarily have to cost a lot of money*" but it is more about "*creating amazing experience.*" While price is an important factor differentiating luxury travel products from conventional travel products, it is not necessarily the only factor. Agents make an effort to provide local and authentic experience: "*we would like to have travel guide who is a local there and who can speak Chinese*" (agent #4). The service quality is high and often predicable. The time poor travelers would like to "*get exactly what they want and make sure it happens*" (agent #3). With the secured high service quality, the experience is considered as value for money even if it is expensive. Agents need to justify the high price. "*They spend it (money) in a very smart way*" (agent #3).

The values luxury travelers look for as mentioned by literature are confirmed by travel agents. Short-term values include safety, privacy, hedonism, and comfort. Safety becomes an important consideration when travelers are traveling to a less-visited destination or traveling with kids: "*they don't want to take risk with the children*" (agent #3). Privacy is often a concern, but it may become less an issue if a destination is exclusive, for example, a trip to polar. "*(A celebrity) is also getting along with tour mates very well... there will be only rational and sophisticated communication.*" Luxury travelers would like to "*enjoy themselves*" (agent #3) and have high standard for comfort in terms of "*size of airplane*" and "*model of vehicle*" (agent #10). Luxury travelers want flexibility to make changes, and they have "*the rights to change the plans*" before or during a trip (agent #5).

Four long-term values including status, uniqueness, self-actualization, and social value are mentioned by agents. Some luxury travelers may wish to display their status by showing off their world travel experience (agent #10), choosing every best item among the alternatives, or visiting places their friends have visited. "*This type of travelers would directly tell you that I want everything best...maybe 10% out of all guests*" (agent #5). But, for many travelers, luxury travel is not "*for showing off*" but only because "*they enjoy it*" (agent #3). Uniqueness is a value that travelers look for and that is why some less-visited destination is becoming popular (agent #7). However, agents seem to have different interpretation on "uniqueness." Agent #11 questioned on whether there will be any unique travel experience: "*anything exclusive... is only going to be temporary.*" Other agents who believe in uniqueness highlight the customization nature of luxury travel experience: "*it is tailor made for you but not copied from elsewhere, so this trip for you is unique*" (agent #4).

Self-actualization is associated with physical challenges during a trip, for example, "*swimming in south polar,*" "*Gobi Challenge*" (agent #11), and "*a trip to Kilimanjaro*" (agent #9). Social value can be accomplished during or after a trip. During the trip, some luxury travelers would like to treat their friends with well-arranged travel activities because "*there will a face concern*" (agent #7) or would like to make friends who share the similar interests because the new friends "*can be helpful in their life and career development*" (agent #9). After a trip, some travelers would share their experience via social media such as WeChat moments among friends. "*They share within a private group...they share because they love the experience, not to show off*" (agent #3).

According to the agents, luxury travel is different from conventional travel product for its better service, uniqueness or customization, memorable and enjoyable experience, exclusiveness, expensiveness, slow pace, and relaxation. The service is better in terms of one-on-one contact, being always connected even after office hour, the ability to provide professional information, and one-stop service on different destinations. "*The requirement on professionalism on luxury travel is much higher than regular tour groups*" (agent #7), for example, agents need to tell "*whether certain flight provides flat-bed-seats in business class*" (agent #6). One agent highlighted the importance of customization: "*I would believe that high-end travel means that we tailor made a tour for him or her*" (agent #11).

Interestingly, although top facilities including luxury hotels, business class flight, and luxury vehicle at the destination are often used by luxury travelers, agents do not describe it as a determinant of luxury travel.

> *Luxury is not about five-star or whatever the price...it's all about the experience. Regardless of whether you stay in a five-star hotel or not, it is a luxury travel because you are in a very unique destination. It's very much who you travel with, where you go, and what kind of experience you are enjoying.*
> (agent #3)

The word "experience" is often used but not clearly defined in luxury travel. According to the agents, luxury experience refers to the memorable and enjoyable activities that the travelers have participated in. For example, "*in Japan, maybe it's tea ceremony*" (agent #10), "*drinking a cup of coffee in a café at Barcelona*" (agent #8), and "*picking vegetables in a garden and cooking with the chef*" (agent #4).

The luxury travel experience is more expensive but also more exclusive. Some experience can only be accessed with high cost such as riding in a Bentley or Rolls Royce (agent #7). Two participants mentioned slow pace as an important feature of luxury travel, which will potentially cost more for a traveler. Travelers are able to spend long enough time at a destination to experience the local life there:

> *We had a customer who visited France with us for three times and spent more than two weeks every time. One time, he even stayed there for almost two months...Being slow is to slow feel the local culture...looking at how locals are shopping and cooking, taking kids to school.*
> (agent #4)

Being able to spare more time at a destination constitutes luxury travel.

Luxury travel agents highlight the importance of service especially their tailor-made itinerary than facilities probably because service is often their competitive advantage. Their understanding of Chinese luxury traveler largely confirmed what is revealed by existing literature but with more variance, for example, the need of privacy varies when travelers join different tours. Their definition of luxury travel or luxury travelers may reflect what they specialize at and their target customers, for example, including high-income customers. Their understanding reflects more the characteristics of their clients who are travelers outsourcing planning to agent. Include a perspective of travelers who may not use agent in the third set of data may enrich the understanding.

Chinese travelers

The travelers who participated in this study can all afford luxury travel products; however, not all of them have the sufficient finance for their ideal way of travel. Some travelers are typical luxury travelers described by the agents or literature. For example, traveler #5 spent about one million RMB (154,000 USD) on leisure trips annually and define luxury travel as: "*You will be able to travel as the way you like without any concern.*" However, some travelers may "*choose the best products considering financial ability*" (traveler #27). Traveler #26 currently always flies in economy class to the destination and first class for the return journey, but would like to always fly in first class if she has more than ten million RMB (US$1,540,000) annual income. In terms of their freedom of using time, 13 travelers consider themselves as time poor. Traveler #7 was an extreme case: "*I and my husband are very busy at work...(during flight) kids fall asleep... we can continue the work.*" Other travelers either have flexible working time or do not work, thus, would have more free time to travel. The understanding from luxury agents has reflected the same status.

As expected by agents, one major reason why travelers would like to have professional travel planning service is their limited knowledge and ability. The middle-aged travelers were not comfortable with planning a trip by themselves and would like to either use agent or rely on their children. Most younger travelers, however, will hesitate to use a travel advisor because they enjoyed the planning process or they do not trust the agent. Traveler #7 found a package online including a business class ticket, but the airline was not the best option between these two cities, thus she lost confidence on the agent. The online-displayed products screen out those travelers who have higher standards. Traveler #20 commented: "*The biggest problem with travel agent is that they always think about how to save money for you...we don't need to save money.*"

Travel products purchased are similar to what was described by agents such as luxury hotels, business class flights, private vehicle with driver, or rent a car for self-driving. They may enjoy fine-dining such as Michelin restaurants (travelers #18) but also enjoy local food. Those who travel with kids may not be able to visit fine-dining restaurants since some restaurants do not accept kids. Travelers may book helicopter or yacht in destinations such as Sanya, China. While many travelers enjoy the convenience of having a private vehicle with driver during the trip, some travelers do not feel comfortable with this service. Traveler #12 mentioned: "*It feels like you are bothering others (the driver) if you had an original plan but would like to visit another place spontaneously. Even if you have paid for this.*"

Travelers' activities vary according to personal interests and travel companion. A few travelers expressed their willingness to visit polar area, but had concern on physical challenge or safety. Traveler #9 mentioned: "*I'd like to stay comfortable during this trip...I know the environmental is harsh.*" Traveler #10 had experienced an accident on a helicopter and would not take helicopter in the future. Hotel is an important part of travel for most travelers, for example, traveler #17 "*would visit a destination for a hotel... If there is no special or nice hotel, I may not visit this destination.*" Traveler

#5 enjoyed a resort in Xishuangbanna, Yunnan, and visited it once a year. Shopping is an important activity, but it depends on destinations. Traveler #2 commented: "*It is not necessary when you visit places like Philippines. But if you visit Europe, you would must consider shopping.*" A few travelers have different preferences when they travel alone, with friends, or with family. For example, Travelers #17 and #19 would like to have adventurous trips when they travel without family.

Travelers also recognized the high price as a characteristic of luxury travel. Traveler #20 preferred paying a price which includes extra services. Price is considered as an indicator of service quality, and travelers would often prefer to choose from the expensive products. A typical practice of these travelers is ranking the hotels by price and choosing from the top ones. Most of them choose expensive products because they would like to have service guarantee. Traveler #1 picked international chain hotel brands because she did not want to have any "*scaring experience.*" Choosing familiar brands also means that their decision-making is easier. Traveler # 3 often chooses the same hotel brand: "*it saves my time and effort in decision making.*"

Travelers may be less price-sensitive, but they would hope to see the value and be charged with a fair price. Traveler #20 will be disappointed if "*you charge others for 1,000 but charge me for 1,500.*" Most travelers are experienced travelers, and they can recognize and appreciate the value, for example, good location of a hotel, tentative service. For example, Traveler #22 appreciates the dining service during flight: "*the tableware…they are all customized…the chinaware is Nikko, a Japanese brand.*" As commented by agents, travelers are spending money in a smart way. Traveler #1 tends to book a better hotel if she plans to enjoy the hotel service and facilities: "*if you consider the time you spend in a hotel and work out the charge per hour, then the rate is lower if you stay in a hotel for a longer period of time.*" Travelers do appreciate local experience and opportunities to understand local culture. For example, Traveler #26 mentioned that: "*I may also want to take bus…the bus in some countries are nice…I would like to be part of the local people.*"

The important short-term values travelers look for include comfort, relaxation, cleanliness, quietness, hedonism, convenience, safety, freedom, and efficiency. Travelers are willing to pay for more comfortable bedding, spacious seating area, high-end vehicle, and better service. Traveler #23 consider travel as a "*tiring activity*" and "*booking a nice hotel will make it more very comfortable.*" Hedonism during a trip is important. Traveler #26 mentioned that: "*It makes me feel happy and I would be able to work better in the future.*" Travelers look for more convenient alternatives. For example, when travelers have a private vehicle, they have less worry about how to get to the place and where to put the personal belongings. Safety is another consideration recognized by agent and confirmed by travelers, for example, luxury hotel may provide "*many security guys, and people will only be able to go upstairs using key card* (traveler #18)." The freedom to make choices is appreciated by travelers. So that the activities will "*match your mood*" (traveler #4).

Relax, cleanliness, and quietness seem less recognized by agent. It is possible that the travelers who simply would like to enjoy quiet moment and to relax would not need agent's planning service. Island destination becomes popular because it is more relaxing compared to cities. Many travelers understand local conditions and would appreciate a clean accommodation at a destination. Traveler #17 described a memorable experience: "*Gobi, desert... Within 20 kilometers, there is nothing but this hotel... But when you go inside, you will find that bath towels are clean, bed is clean, not extravagant...that's enough.*" The requirement on quiet environment is high among these travelers. Traveler #6 would prefer business class or first class because he needs quiet environment for reading or thinking. Choosing luxury travel product also potentially save travelers' time. For example, luxury hotels often are located in good locations and save transportation time.

While agents have a good understanding on travelers' short acting values, the long-lasting values revealed by travelers themselves are more diverse than agents' expectation. A few travelers consider status as value they would like to achieve. Traveler #5 felt that service provided by luxury hotel has satisfied her "*vanity.*" Novelty is more important for travelers, and uniqueness is less important. They would like to have new experience they never had. Social value can be observed. Traveler #19 considered a by-invitation tour as a network. Traveler #11 described travel as a reunion with old friends. The most significant long-lasting value for these travelers is horizon broadening or learning. Traveler #15 believed that the impression of a place may change after a visit and that is what "*travel can bring to you.*" Traveler #22 emphasized the importance of being at a place physically: "*if you see some images in a movie, you may forget after five or ten years. But if you have touched and experienced...you can remember that for your whole life.*" A few travelers also mentioned a sense of achievement because not everyone can have the same travel experience.

The difference between luxury travel and conventional travel described by travelers shows similar characteristics mentioned by agents. Travelers consider these products as expensive, but are ready to pay the premium for better service and memorable, steady, and relaxing experience. For example, Traveler #1 enjoyed the butler service in hotel: "*We will not like to find a new person when we have a new request. We all enjoy looking for the same person.*" Luxury experience such as distinctive luxury hotel may bring them a memorable experience. For example, Traveler #9 remembers the details of decoration in a luxury hotel: "*when you try to recall, the room is still in your mind.*" Even though they are time poor and would like to utilize the time during the trip, travelers would enjoy a slow pace during travel: "*the perfect travel should be comfortable, and the pace is not too fast. And you can enjoy the time with family.*" Customization is only mentioned by travelers who use agent. Traveler #20 described one satisfying experience where the tour guide was able to find a restaurant where smoking is allowed for him.

Conclusion

The findings suggest the following six common themes about "Chinese luxury travelers." Using these themes, researchers or practitioners may clarify their understanding when discussing this segment. The meaning of "Chinese luxury travelers" often consists of a minimum level of financial support either from their own pocket or from companies (Wealth), a general preference toward luxury facilities (Facilities) and high-quality service (Service Quality), and an expectation of exclusive, authentic, and slow destination experience (Destination Experience). The preference of high-end facilities and service is triggered by their desire for pleasure as short-term value (Pleasure-Seeking) and various long-term values such as status and learning depending on their daily life (Contrast).

Wealth or financial status of luxury travelers is understandably important. Different levels of financial status have been used to describe Chinese luxury travelers. The highest level is that travelers would be able to purchase any travel products without any concern on budget. This criterion will exclude many travelers who often stay in luxury hotels but travel in economy class for short haul to contain budget. A loose criterion is that travelers at least have savings to complete one "expensive" trip, for example, once-in-a-lifetime visit to the Arctic. Often time, a definition will fall between these two extreme ends, and Chinese luxury travelers can be considered as a group of travelers who have sufficient financial resources for their regular consumption of travel products at above-average prices.

The theme about facilities concerns hotels, flights, and other transportation means. Luxury travelers are characterized by the use of luxury hotels, first class or business class flights, luxury cruises, limousines with chauffeurs, helicopters, and leisure yachts. Depending on the type of travel experience that luxury travel agents provide and what travelers prefer, their use of such facilities varies. Some agents would provide their service only if their customers are willing to pay for such facilities. On the other hand, some travelers would visit remote and exotic places even if there is no luxury accommodation. In general, Chinese luxury travelers consider that the availability of luxury facilities is important to ensure remarkable experience.

High-service quality is expected in luxury travel. Travelers are concerned with the much better service of luxury travel compared to the conventional travel. Luxury travelers expect high standards of service quality and are willing to pay a premium for it. Dissatisfaction easily appears when their expectation is not met. There are, however, exceptions when some travelers may visit remote places where they have to cook and prepare tent by themselves, for example, when traveling in a caravan or having adventurous trips. In general, Chinese luxury travelers look for high-service quality from every service provider they would encounter during a trip, and professional luxury travel agents can help steer them in the right direction.

Luxury travelers tend to look for destination experience which is exclusive, authentic, and slow. They value exclusive experience such as the breath-taking ocean view at a luxury hotel, authentic cooking lesson with a local chef, taking the time to explore a city slowly, and enjoying a birthday party with the beloved ones. The experience can be potentially be provided by service providers, however, is often only performed by luxury travel service providers. For example, luxury hotels would often celebrate birthday for guests. In general, Chinese luxury travelers can be considered to value a destination experience a lot if they would regret not visiting the place and they wish to visit the place again if there is such an opportunity.

The pleasure-seeking theme is concerned with the motivation of luxury travelers. They consider pleasure as an important short-term value they look for during trips, thus would like to both maximize pleasure such as enjoying fine-dining food and minimize the tiredness and effort often associated with transportation such as hiring a driver with a high-end vehicle. This explains why they may spend more on travel compared to other people who have a similar wealth status. The attributes linked to this theme include comfort, relaxation, quietness, hedonism, convenience, and freedom. In general, Chinese luxury travelers are particular about the level of pleasure experienced in their travel, and it is important for the reward to be noticeable.

When travel is just an extension of normal life, travelers may simply look for pleasure and some basic needs in an unfamiliar place. In luxury travel, people have much higher expectation. They may wish to learn something from the travel experience, they may want to show off their status, they may want to indulge themselves during the trip, and they may see the experience as a reward for their hard work or something to please their beloved ones. In general, Chinese luxury travelers tend to look for contrast to their normal life. Some travelers would like to enjoy luxury temporarily during the trip. For those who have already achieved high social status and enjoy daily luxury, they look more for self-fulfillment and sense of elevation.

Who are Chinese luxury travelers? They are spending on and enjoying top facilities and services provided by the travel industry and are still exploring new destinations and elevating their experience with their growing wealth and knowledge on travel. They are experienced travelers looking for pleasure during trip and various long-lasting personal values from destination experience. They appreciate the high-quality service currently available in tourism industry and are also expecting the industry to understand their needs better. Should we associate Chinese luxury travelers with Aman Resorts and Michelin restaurants? May be. They are also shaping luxury travel for tomorrow by their own travel experiences.

References

Amadeus. (2016). *Shaping the future of luxury travel: Future traveler tribes 2030*. Retrieved April 15, 2021, from http://www.amadeus.com/documents/future-traveler-tribes-2030/luxury-travel/shaping-the-future-of-luxury-travel-report.pdf

Bakker, M. (2005). *Luxury and tailor-made holidays, travel and tourism analyst*. Chicago, IL: Mintel International Group.

Bowerman, G. (2014). *The new Chinese traveler: Business opportunities from the Chinese travel revolution*. London: Palgrave Macmillan. doi:10.1057/9781137397331_3

Brenner, L., & Aguilar, A. G. (2002). Luxury tourism and regional economic development in Mexico. *The Professional Geographer*, *54*(4), 500–520. doi:10.1111/0033–0124.00346

Chen, N., Masiero, L., & Hsu, C. H. C. (2019). Chinese outbound tourist preferences for all-inclusive group package tours: A latent class choice model. *Journal of Travel Research*, *58*(6), 916–931. doi:10.1177/0047287518794331

Correia, A., Kozak, M., & Del Chiappa, G. (2019). Examining the meaning of luxury in tourism: A mixed-method approach. *Current Issues in Tourism*, *23*(8), 952–970. doi:10.1080/13683500.2019.1574290

Fairmont Hotels & Recorts. (2018). *New research study from fairmont shows affluent travellers want exclusive experiences and view hotels as trusted travel partners*. Retrieved April 15, 2021, from https://www.fairmont.com/promo/luxuryreports/luxury-insights-reports/pdfs/gateway-to-home.pdf

Globetrender. (2020). *China next: The future of luxury tourism*. Retrieved April 15, 2021, from https://globetrender.com/downloads/china-next-the-future-of-luxury-tourism/

HH Travel. (2016). *The report of luxury travel trend in China by HH Travel (2016 spring-summer)*. Retrieved April 15, 2021, from https://m.ctrip.com/html5/you/news/list-ctrip/20513.html

Horwath HTL. (2018). *Rethinking luxury in hospitality*. Retrieved April 15, 2021, from https://corporate.cms-horwathhtl.com/wp-content/uploads/sites/2/2018/05/HHT_Rethinking-Luxury.pdf

Hung, K., Huang, H., & Lyu, J. (2020). The means and ends of luxury value creation in cruise tourism: The case of Chinese tourists. *Journal of Hospitality and Tourism Management*, *44*, 143–151. doi:10.1016/j.jhtm.2020.05.004

Hurun Report. (2018). *Hurun Report released the Chinese luxury traveler 2018 report*. Retrieved April 15, 2021, from http://www.hurun.net/EN/Article/Details?num=A405F6502C2D

Iloranta, R. (2019). Luxury tourism service provision-lessons from the industry. *Tourism Management Perspectives*, *32*, 100568. doi:10.1016/j.tmp.2019.100568

International Luxury Travel Market. (2019). *The rise and rise of the Chinese luxury travelers and their spending power*. Retrieved April 15, 2021, from https://www.iltm.com/en/china/News/Press-Releases/THE-RISE-AND-RISE-OF-THE-CHINESE-LUXURY-TRAVELLERAND-THEIR-SPENDING-POWER/

Jing Culture & Commerce. (2020). *What do Chinese luxury travelers want? Adventure, of course*. Retrieved April 15, 2021, from https://jingculturecommerce.com/what-do-chinese-travelers-want-experience/

MeetingsNet. (2006, December 18). *Redefining the high end: Ritz-Carlton's luxury meetings forum*. Retrieved April 15, 2021, from http://www.meetingsnet.com/news/redefining-high-end-ritz-carltons-luxury-meetings-forum

Novotná, M., & Kunc, J. (2019). Luxury tourists and their preferences: Perspectives in the Czech Republic. *Tourism: An International Interdisciplinary Journal*, *67*(1), 90–95. https://hrcak.srce.hr/file/318812

Ovation Travel. (2018). *From the front line: Luxury vacation trends 2018*. Retrieved April 15, 2021, from https://www.ovationtravel.com/luxury-trends-2018

Popescu, I. V., & Olteanu, V. (2014). Luxury tourism: Characteristics andtrends of the behavior of purchase. *SEA - Practical Application of Science*, *2*(4), 319–324. http://seaopenresearch.eu/Journals/articles/SPAS_4_35.pdf

Resonance Consultancy Ltd. (2016). In *2016 the future of Chinese International Travel*. Retrieved April 15, 2021, from https://www.discoverlosangeles.com/sites/default/files/media/B2B/RES-Portrait-Chinese-International-Traveler-5.pdf

Savin-Baden, M., & Major, C. H. (2013). *Qualitative research: The essential guide to theory and practice*. Abingdon: Routledge.

Skift. (2017). *Building brand love and loyalty in luxury hospitality*. Retrieved April 15, 2021, from https://view.iltm.com/wp-content/uploads/2017/05/ILTM-Apr20-final.pdf

Stankova, M., & Kaleichev, S. (2013). The cutting edge between the mass and luxury tourism. *Romanian Economic Business Review*, *8*(4), 50–59. http://www.rebe.rau.ro/REBE-WI13S.pdf#page=50

Statista. (2019). *Number of China's high-net-worth individuals (HNWI) with investable assets from 2008 to 2019*. Retrieved April 15, 2021, from https://www.statista.com/statistics/1013630/china-number-of-hnwis-with-investable-assets/

Swarbrooke, J. (2018). *The meaning of luxury in tourism, hospitality and events*. Oxford: Goodfellow Publishers.

Virtuoso. (2019). *The 2020 Virtuoso Luxe Report*. Retrieved April 15, 2021, from https://www.virtuoso.com/getmedia/4f60d91e-e78d-464f-89f1-65d666baf488/2020-Virtuoso-Luxe-Report-Release_FINAL.aspx

Wu, B., & Yang, W. (2018). What do Chinese consumers want? A value framework for luxury hotels in China. *International Journal of Contemporary Hospitality Management*, *30*(4), 2037–2055. doi:10.1108/IJCHM-08–2016–0466

Zhang, E. Y., & Tse, T. S. M. (2018). Tapping into Chinese Luxury Travelers. *Journal of China Tourism Research*, *14*(1), 71–99. doi:10.1080/19388160.2018.1437102

16 Effects of COVID-19 knowledge on Chinese tourists' intentions for medical purposes

Foad Irani, Sina Vahedi and Sima Rahimizhian

Introduction

When the year 2019 came to an end, the most dreadful pandemic began. Any epidemic, such as COVID-19, has created an uncertain atmosphere for businesses and industries, especially the tourism industry. Self-protective behavior among people facing lethal risks, such as pandemics, has risen due to this cause. During the pandemic, agitation and fear of virus infectivity resulted in this self-protective action (Chuo, 2014; Sparks & Pan, 2009; Vahedi & Irani, 2021). Individuals show more COVID-19-related behavioral symptoms of anxiety, which significantly affects their mental health (Mertens et al., 2020; Reisinger & Mavondo, 2005). According to the 2020 report of the United Nations World Tourism Organization (UNWTO), stress and tension could be visible in downward tourism patterns. The UNWTO also reported a 70% decrease in global travel patterns. Thus, in the wake of the COVID-19 pandemic, countries with a higher population would be more impacted by the tourism decline than others. This issue has resulted in perceived risks (PR) and obstacles for travelers all over the world.

Accordingly, this protective behavior made other countries follow a prevention policy toward Chinese tourists because world assemblies assumed responsibility for the virus' spread worldwide. Their lifestyles and eating habits had amplified this into a disaster (Campbell, 2020; Rich, 2020). This downward trend in tourism also affects the psychological risk and impacts the person's intention to travel as well as medical tourism. The medical tourism concept applies to the scheduled and arranged leisure trips outside the country for the refreshment and healing of the body and soul of all people, all across the world (Carrera & Bridges, 2006). Medical tourism is motivated mainly by the need to find a higher or greater level of hospitality for medical purposes (Edelheit, 2008). There is no doubt that, by understanding the principle of medical tourism, countries with a larger population would have a greater appetite for medical tourism.

Since 2011, China's mainland republic has opened its doors and described its visitors as "free, independent travelers". This new approach by China's government has blown fresh air into the type and purpose of travelers, and

DOI: 10.4324/9781003121329-19

by having an overview of this trend, Shen et al. (2016) have reported that this incident has made Chinese citizens travel to other countries, especially to Taiwan, which is the nearest and most accessible country for medical purposes. Similarly, Liu and Johnson (2016) showed that Chinese travelers prefer to travel to the United States more than to other countries for medical purposes. Taiwan got the second most Chinese medical tourists after the United States, and medical tourism has found a place among Chinese outbound travelers. Although all the studies have proven the association between perception of risks and the intention of travel, there is a lack of significant policy to overcome the restrictions during and after COVID-19 in the medical tourism context. Thus, this paper aims to investigate the impacts of perceived knowledge (PK) of COVID-19 risk on Chinese medical tourism and their travel intention (TI). This can brighten our vision of their future concerns about TI for medical purposes.

Literature review

Travel to and within several countries is prohibited while the globe continues to battle the COVID-19 outbreak (Rastegar et al., 2021). Crisis management and public relations strategies like vaccines, together with health measures such as these, may help the image recover and be restored after a calamity (Coombs, 2020). As a consequence, it's crucial since the destination's overall image through knowledge perception influences travel intentions to that location, and visitors' perceptions of destination risk impact travel intentions greatly (Chen & Chen, 2010; Irani & Rahimizhian, 2021). As a result of this kind of behavior, people are more likely to travel safely by avoiding any potential hazards (Irani et al., 2021; Rastegar et al., 2021). People are expected to keep away from perceived high-risk areas even if travel is possible during the COVID-19 crisis, making these kinds of concerns crucial (Hotle, 2020). Therefore, knowledge and risk perception shape the Chinese outbound tourist's intention to visit a medical destination.

Theory of planned behavior (TPB)

According to the TPB, every individual's behavioral intention is constructed by their attitude toward social norms, community and behaviors (Ajzen, 1985; Quintal et al., 2010). This theory has expanded over time, and many researchers have used it in different disciplines (Armitage & Conner, 2001; Steinmetz et al., 2016). Three variables, according to TPB, influence intentions (Ajzen, 1985). First, the attitude toward behavior is a positive or negative evaluation of an individual. Second, subjective norms are described as the social obligation and perceived pressure used by others on an individual to engage in or abstain from a particular behavior. Behavioral control is the final determinant of whether you feel at ease or difficult performing a specific action based on previous personal experience and the required limitations. In line with the TPB model, this research presumes that people behave in a particular way if they have a specific and advantageous outcome. They will be more concerned when they are convinced that the essential resources, capacities, and possibilities are available to act in a certain way (Irani et al., 2021).

Perceived knowledge and perceived risks

Self-reported knowledge impressions are used to measure people's perceptions of how much knowledge they have about a topic (Wallace et al., 2019). PR is characterized by complexity and consequence, with higher levels of confusion and the likelihood of more negative outcomes increasing PR (Oglethorpe & Monroe, 1987; Tavitiyaman & Qu, 2013). The adverse effect of perceived travel risk from the host's perspective as a result of knowledge perception toward COVID-19 has been explored in recent studies (Armutlu et al., 2020; Hadi et al., 2020; Hadood & Irani, 2020), and the attitude of the host community creates the tourism environment (do Valle et al., 2012).

The hospitable behavior influenced by the perception of knowledge regarding COVID-19 influenced risk perception among the host community (Armutlu et al., 2020; Campbell, 2020; Choi et al., 2013; Dash, 2021; Liang et al., 2019; Rich, 2020). Likewise, the impact of COVID-19 on the growing sense of bigotry and xenophobia was linked to Chinese tourists ("Sinophobia: How a virus reveals the many ways China is feared," 2020). According to prior studies based on TPB, the relationship between the PK of tourists and their intention to travel due to PR has been discussed (Han et al., 2020; Khan et al., 2020; Sánchez-Cañizares et al., 2020). Therefore, the following hypotheses have developed:

H1: The PK of COVID-19 negatively influences PR.

Perceived risks and tourist's travel intention

Certain events such as natural disasters, violence, extremism, and epidemics would dramatically raise the PR of any person traveling to their favorite destination (Chew & Jahari, 2014; Fuchs & Reichel, 2006; Rahimizhian & Irani, 2020; Rittichainuwat & Chakraborty, 2009). Thus, based on prior studies, the importance of the PR-TI collaboration was emphasized (Lam & Hsu, 2006; Li & Ito, 2021; Perić et al., 2021; Rather, 2021). For this issue, it is critical to examine the previous relationship during the COVID-19 pandemic. Based on the literature review, risk perceptions harmed intentions to travel among tourists (Li & Ito, 2021; Neuburger & Egger, 2020). For example, the significant impact of PR on TI on the behavioral traits of tourists in China and Japan was investigated (Li & Ito, 2021). Neuburger and Egger (2020) have also discussed a similar study with travelers in the DACH area (e.g., Germany, Austria, and Switzerland). They used different terms to describe

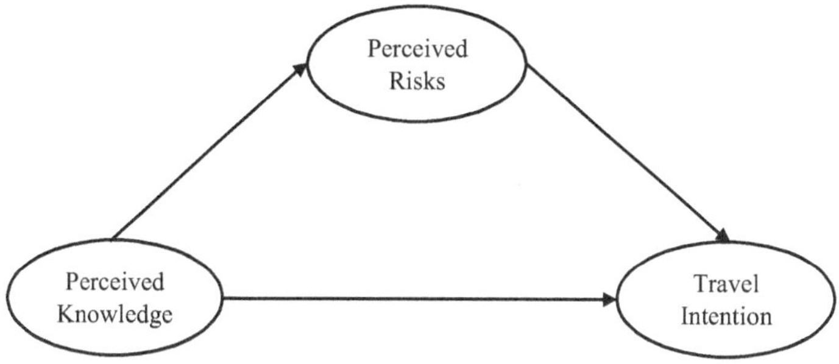

Figure 16.1 Research model of the study

PR and TI. Therefore, the following hypothesis was developed based on the relevant context:

H2: The PK of COVID-19 negatively influences TI.

Mediation role of perceived risks

Based on extant literature, the role of PR has had a significant effect on medical tourism in different destinations worldwide (Chen et al., 2020; Seow et al., 2017; Wang, 2012). Likewise, Seow et al. (2017) illustrated the effectiveness of tourists' PR on attitudes and intentions to revisit Malaysia. Similarly, Wang (2012) explored PR's effects on perspective and TI of Chinese outbound tourism for rehabilitation. For instance, Tavitiyaman and Qu (2013) investigated PR as a mediator of the association between destination image and behavioral intention based on the TPB framework. During the COVID-19 pandemic, former studies have comprehended the TPB framework with pandemic-related topics (Han et al., 2020), specifically the direct link between PR and the intention of seeking a safer destination to travel to (Khan et al., 2020; Sánchez-Cañizares et al., 2020). Thus, based on the literature review, Chinese outbound has received less attention in the medical tourism context (Figure 16.1). Therefore, based on the literature review, the following hypothesis was developed:

H3: PR mediates the association between PK and tourist travel intention.

Methodology

The proposed model and the hypotheses developed by this study were estimated in a survey questionnaire. The data were collected from Chinese outbound tourists who traveled to Turkey for medical purposes by utilizing a judgmental sampling method. In recent years, the global market has seen health and wellness tourism develop rapidly. Turkey also spends so much money to help demand progress for long-term plans, making Istanbul strenuously at the top. Over 20 million people travel the globe for medical treatment and surgery. The US, South Korea, Turkey, Thailand, Germany, India, the United Kingdom, Malaysia, Mexico, and Iran are the ten most visited nations for their accredited hospitals, quality of services, and affordable healthcare.

Turkey has ranked in the top ten countries with state-of-the-art hospitals, quality, affordable healthcare, spa treatments, natural hot spring water supplies, and renowned global hospitality in recent years, attracting numerous medical tourists. Over two million medical tourists worldwide visit Turkey for health tourism, and Istanbul is the first stop for those traveling to Turkey for medical and health tourism, especially those from Europe, Asia, and the Middle East. By emphasizing the purpose of this study, the research team received consent from the Directors of Cosmetic, Esthetic, and Hair Transplant Operations to collect data. Thus, both directors promised to collect data from their Chinese clients.

Statistical remedies were applied to minimize the possibility of common method bias (CMB) (Podsakoff et al., 2003). The research team guaranteed respondents' anonymity and were asked to participate voluntarily (Hair et al., 2014). The incomplete questionnaires were discarded for further consideration. A total of 340 questionnaires were distributed among participants. After discarding the incomplete copies, 322 (93.6%) completed questionnaires were retrieved. The majority of the participants were males, 198 (61.5%), and the remaining were females. Sixty-two percent of the respondents were aged between 28 and 37 years. Almost half of the respondents (47.8%) had a bachelor's degree. Regarding the monthly income, 36% of the respondents had an income of 1,500 to 2,000 US dollars monthly.

The scale items were adopted from the extant literature. Five-point Likert scale ranging from 5-Strongly Disagree to 1-Strongly Agree was used to measure the scale items. The PR was measured by adapting three items developed by Khan et al. (2020). The three items were used to measure the PK of COVID-19 by Han et al. (2020) and Sánchez-Cañizares et al. (2020). Medical tourists' TI was measured by adapting three items developed by Hsu and Huang (2010). The original version of the questionnaire was designed in English and then translated into Chinese by utilizing the back-translation method. The back-translation was done by two independent experts who are fluent in English and Chinese (Chen & Boore, 2010). Furthermore, the pilot study was applied by distributing the questionnaire among 20 Chinese participants to control the comprehensibility scale items.

SPSS version 24 was used for descriptive statistical analysis, and the hypotheses were examined by applying ADANCO version of 2.1.1 (Ringle et al., 2015). The common method variance (CMV) was determined as 47.4% using Harman's single factor evaluation. Thus, the value of a single factor revealed that CMB, statistically, was not an issue due to having a value less than the threshold of 50% (Kock, 2015).

Results

ADANCO 2.1.1 was used to conduct the confirmatory factor analysis (CFA). The data reliability was confirmed by assessing the indicator loadings, Dijkstra-Henseler's ρ (ρA), Jöreskog's ρ (ρC), Cronbach's alpha (α), Average Variance Extracted (AVE), and Composite Reliability (CR) as shown in Table 16.1 (Dijkstra & Henseler, 2015). Regarding internal consistency and reliability, CR should be greater than 0.7 (Hair, 2006); thus, all constructs in this study met the condition as shown in Table 16.1. The indicators are loaded under their underlying constructs. Hence, all indicator loadings of this study's constructs ranged from 0.708 to 0.899, which shows the primary evidence of convergent validity. Additionally, constructs' AVE value should be greater than 0.5, and the constructs in this study passed this condition.

The discrimination validity has been assessed by using the Fornell and Larcker (1981) and Heterotrait-Monotrait ratio criterion. Table 16.2 represents the results that confirmed the discrimination validity of all study constructs, as all the constructs' AVE had a greater value than the squared correlation based on Fornell and Larcker's (1981) demand that every inter-correlation value has a Heterotrait–Monotrait ratio of less than 0.9.

Results of hypothesized relationship

Our first hypothesis proposes a direct link that PK affects PR. The results show that PK has a significant influence on PK (= 0.52, 9.29**), so Hypothesis 1 received significant support. Hypothesis 2 proposes that PR significantly

Table 16.1 Results of confirmatorly factor analysis

Items	Loadings	(ρA)	(ρC)	(α)	AVE	CR
Perceived knowledge		**0.85**	**0.87**	**0.78**	**0.69**	**0.87**
PK1	0.876					
PK2	0.899					
PK3	0.708					
Perceived risks		**0.82**	**0.81**	**0.73**	**0.89**	**0.89**
PR1	0.888					
PR2	0.833					
PR3	0.839					
Travel intention		**0.85**	**0.91**	**0.85**	**0.77**	**0.91**
TI1	0.865					
TI2	0.895					
TI3	0.865					

Note: ρA implies Dijkstra–Henseler's rho, ρC implies Jöreskog's rho, AVE implies average variance extracted, α implies Cronbach's alpha, and CR is composite reliability
Source: Authors' clarification

Table 16.2 Discriminant validity results

Construct	PK	PR	TI
HTMT ratio criterion			
PK			
PR	0.6170		
TI	0.5406	0.5414	
Items	**PK**	**PR**	**TI**
Fornell and Larcker criterion			
PK	**0.6918**		
PR	0.2662	**0.7290**	
TI	0.2297	0.2117	**0.7661**

Note: Square root of AVE (in bold on diagonal) and constructs correlation coefficients, PK is perceived knowledge, PR implies perceived risks, and TI represents tourists' travel intention
Source: Authors' clarification

Table 16.3 Result of direct, indirect and total effects

Effect	Beta	Indirec effect	Total effect	t-Value	Cohen's f^2	R^2	Remark
PK->PR	0.52		0.52	9.29^{**}	0.36		Supported
PK → TI	0.33	0.15	0.48	10.42^{**}	0.11		Supported
PR → TI	0.29		0.29	5.08^{**}	0.08		Supported
PR						0.26	
TI						0.29	

Note: resample size for each Bootstrapped: 9.999, PK implies perceived knowledge, PR implies perceived risks, and TI implies tourists' travel intention. **$P \leq 0.05$
Source: Authors' elaboration

influences tourists' TI. The results show that PR has a significant effect on TI (= 0.29, 5.08**), so H2 is also supported (Table 16.3).

Finally, we proposed Hypothesis 3: that PR mediates the association between PK and tourists' TI. The mediation effect in ADANCO is evaluated using Preacher and Hayes's (2008) two-step approach. In the first step, we perform a bootstrapping analysis to check the significance of the direct relationship without mediating variables (e.g., PR). Furthermore, the same approach was repeated with a mediating variable to confirm the indirect effect significance. The mediating effect of PR on PK and tourists' TI was confirmed since the indirect effect was significant. Hence, H3 received support. This result shows that PK, together with PR, influences tourists' TI (Figure 16.2).

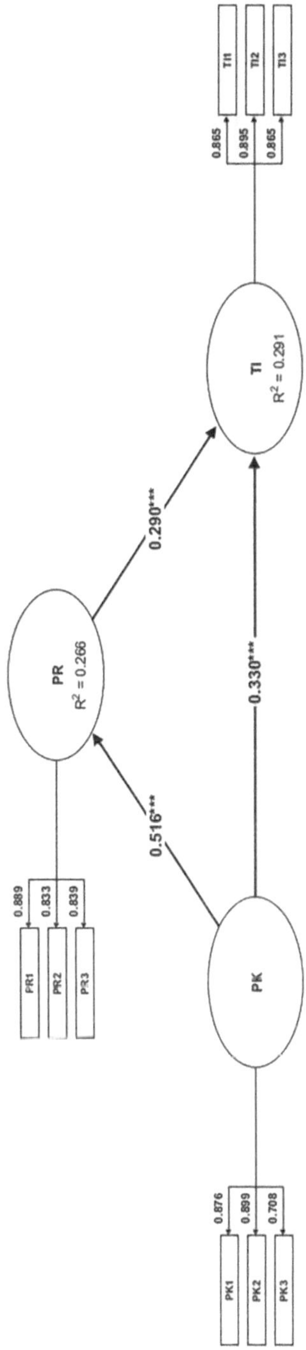

Figure 16.2 Measurement of research model

Conclusion

This study is significant because it raises awareness among people in the travel industry about how discrimination affects the health and lifestyle of Chinese citizens around the world, including those who travel to Turkey for medical purposes. This study examined the impact of PK on PR and TI of Chinese outbound medical tourism during the pandemic period. According to the results, the PK of COVID-19 significantly affects Chinese tourists' PR and TI. This is in line with the "Sinophobia: How a virus reveals the many ways China is feared" (2020) conclusion on the host community's mindset toward Chinese outbound. Consequently, the unfamiliar environment for Chinese outbound medical tourists will build a sense of conservatism for this nation's short-term medical travel.

This research broadens our understanding of PK and PR that influence the Chinese tourist's TI toward medical tourism in Turkey from a theoretical point of view. The development of the TPB was one of the main theoretical implications of this study. Moreover, the study's findings have shown the importance of PK and risks in exploring the Chinese tourist's TI to purchase medical services in Turkey. Previous studies have rarely investigated PK and PR's effects on tourists' TI in the medical tourism context. Our hypotheses and investigation results contribute to existing knowledge in a variety of ways.

First, it expanded the impact of PK of COVID-19 in the literature. Although this relationship has been discussed in Han et al. (2020), the effect of PK on TI based on the TPB framework has not been examined. Therefore, this study expanded the previous TPB framework by examining new relations among variables. Second, PR and TI have been widely used in literature to mediate the role of attitude based on the TPB framework (Dash, 2021; Han et al., 2020). Prior research has not been conducted on the direct impact of PR on TI. Therefore, this study provided a theoretical model based on the TPB framework to measure PR's effectiveness on TI for Chinese outbound medical tourists. Finally, PK was analyzed by Han et al. (2020), and this research added a new viewpoint to the literature on PK in the sense of medical tourism. Han et al. (2020) examined PR as one of the variables in their models. However, they have not studied the direct effect of PK on PR. Therefore, due to the gap in the relevant literature, this study investigated the effect of PK on PR.

There are some managerial implications for this study. First, management can educate themselves on COVID-19 and its social and psychological impact on the host population. They should include appropriate guidance for institutes and recommend similar principles to their constitutional organization to raise consciousness among their community to provide the same medical facilities to Chinese outbound visitors to reduce prejudice among travel industry stakeholders. Second, they should have special facilities for those traveling from the Republic of China to satisfy their needs before joining civilization for the locals' sake to avoid the unwelcome dissemination of the COVID-19 virus. As a result, Turkish governors should provide tourism institutes with PCR tests and percussive equipment for hotel and travel industry employees to prevent the spread of possible pathogens.

This study has limitations, and therefore, further recommendations are needed to ensure the results. First, we have only investigated this trend in Turkey. Consequently, future studies advise considering other destinations to improve the result's generalizability. We recommend creating new lines of investigation, such as completing the study to examine different countries or cultures coping with distinct pandemic conditions. Furthermore, the host community's psychological issues were not addressed in this study. Thus, researchers must conduct related studies to identify other potential social, cultural, and economic effects on locals. Finally, it is essential to stress the importance of human health matters in all situations. These solutions will shed light on the challenging pandemic era, demonstrating the value of medical tourism for human rights purposes and reducing unnecessary restrictions for medical tourists, especially Chinese people.

References

Ajzen, I. (1985). From intentions to actions: A theory of planned behavior. In J. Kuhl & J. Beckmann (Eds.), *Action control: From cognition to behavior* (pp. 11–39). Heidelberg, Berlin: Springer.

Armitage, C. J., & Conner, M. (2001). Efficacy of the theory of planned behaviour: A meta-analytic review. *British Journal of Social Psychology, 40*(4), 471–499. doi:10.1348/014466601164939

Armutlu, M. E., Bakır, A. C., Sönmez, H., Zorer, E., & Alvarez, M. D. (2020). Factors affecting intended hospitable behaviour to tourists: Hosting Chinese tourists in a post-COVID-19 world. *Anatolia: An International Journal of Tourism and Hospitality Research, 32*(2), 218–231. doi:10.1080/13032917.2020.1855595

Campbell, C. (2020). The West blames the Wuhan Coronavirus on China's love of eating wild animals. The truth is more complex. *Time Magazine, 24*.

Carrera, P. M., & Bridges, J. F. (2006). Globalization and healthcare: Understanding health and medical tourism. *Expert Review of Pharmacoeconomics & Outcomes Research, 6*(4), 447–454. doi:10.1586/14737167.6.4.447

Chen, C. F., & Chen, F. S. (2010). Experience quality, perceived value, satisfaction and behavioral intentions for heritage tourists. *Tourism Management, 31*(1), 29–35.

Chen, H. Y., & Boore, J. R. (2010). Translation and back-translation in qualitative nursing research: Methodological review. *Journal of Clinical Nursing, 19*(1–2), 234–239.

Chen, X., Cheng, Z. F., & Kim, G. B. (2020). Make it memorable: Tourism experience, fun, recommendation and revisit intentions of Chinese outbound tourists. *Sustainability, 12*(5), 1904.

Chew, E. Y. T., & Jahari, S. A. (2014). Destination image as a mediator between perceived risks and revisit intention: A case of post-disaster Japan. *Tourism Management, 40*, 382–393. doi:10.1016/j.tourman.2013.07.008

Choi, J., Lee, A., & Ok, C. (2013). The effects of consumers' perceived risk and benefit on attitude and behavioral intention: A study of street food. *Journal of Travel & Tourism Marketing, 30*(3), 222–237. doi:10.1080/10548408.2013.774916

Chuo, H.-Y. (2014). Restaurant diners' self-protective behavior in response to an epidemic crisis. *International Journal of Hospitality Management, 38*, 74–83. doi:10.1016/j.ijhm.2014.01.004

Coombs, W. T. (2020). Conceptualizing crisis communication. In R. L. Heath &. H. D. O'Hair (Eds.), *Handbook of risk and crisis communication* (pp. 99–118). New York: Routledge.

Dash, A. (2021). Exploring visit intention to India for medical tourism using an extended theory of planned behaviour. *Journal of Hospitality and Tourism Insights, 4*(4), 418–436. doi:10.1108/jhti-03-2020-0037

Dijkstra, T. K., & Henseler, J. (2015). Consistent partial least squares path modeling. *MIS Quarterly, 39*(2), 297–316. doi:10.25300/MISQ/2015/39.2.02

do Valle, P. O., Mendes, J., & Guerreiro, M. (2012). Residents' participation in events, events image, and destination image: A correspondence analysis. *Journal of Travel & Tourism Marketing, 29*(7), 647–664. doi:10.1080/10548408.2012.720151

Edelheit, J. S. (2008). Defining medical tourism or not. *Medical Tourism Magazine, 5*(1), 9–10.

Fornell, C., & Larcker, D. F. (1981). Structural equation models with unobservable variables and measurement error: Algebra and statistics. *Journal of Marketing Research, 18*(3), 382–388. doi:10.1177/002224378101800313

Fuchs, G., & Reichel, A. (2006). Tourist destination risk perception: The case of Israel. *Journal of Hospitality & Leisure Marketing, 14*(2), 83–108. doi:10.1300/j150v14n02_06

Hadi, D. M., Irani, F., & Gökmenoğlu, K. K. (2020). External determinants of the stock price performance of tourism, travel, and leisure firms: Evidence from the United States. *International Journal of Hospitality & Tourism Administration*, 1–17. doi:10.1080/15256480.2020.1842838

Hadood, A. A. A., & Irani, F. (2020). Impact of economic sentiment and economic policy uncertainty on travel and leisure stock return. *Journal of Hospitality and Tourism Insights*, ahead-of-print. doi:10.1108/jhti-05-2020-0070

Hair Jr., J. F. (2006). Successful strategies for teaching multivariate statistics. In *Proceedings of the 7th International Conference on Teaching Statistics.*

Hair Jr., J. F., Sarstedt, M., Hopkins, L., & G. Kuppelwieser, V. (2014). Partial least squares structural equation modeling (PLS-SEM). *European Business Review, 26*(2), 106–121. doi:10.1108/ebr-10-2013-0128

Han, H., Al-Ansi, A., Chua, B.-L., Tariq, B., Radic, A., & Park, S. (2020). The post-coronavirus world in the international tourism industry: Application of the theory of planned behavior to safer destination choices in the case of US outbound tourism. *International Journal of Environmental Research and Public Health, 17*(18), 6485. doi:10.3390/ijerph17186485

Hotle, S., Murray-Tuite, P., & Singh, K. (2020). Influenza risk perception and travel-related health protection behavior in the US: Insights for the aftermath of the COVID-19 outbreak. *Transportation Research Interdisciplinary Perspectives, 5*, 100127.

Hsu, C. H. C., & Huang, S. (Sam). (2010). An extension of the theory of planned behavior model for tourists. *Journal of Hospitality & Tourism Research, 36*(3), 390–417. doi:10.1177/1096348010390817

Irani, F., Ozturen, A., & Akhshik, A. (2021). Host-tourist interactions and residents' attitudes towards sustainable tourism development. In A. Decrop, A. Correia, A. Fyall, & M. Kozak (Eds.), *Sustainable and collaborative tourism in a digital world* (pp. 123–134). Oxford: Goodfellow Publishers.

Irani, F., & Rahimizhian, S. (2021). Exploring the association among intention to study abroad, destination brand awareness, WOM, and destination choice of international students of North Cyprus. In *Managing Tourism Across Continents (MTCON'21)*.

Khan, M. J., Khan, F., Amin, S., & Chelliah, S. (2020). Perceived risks, travel constraints, and destination perception: A study on Sub-Saharan African medical travellers. *Sustainability*, *12*(7), 2807. doi:10.3390/su12072807

Kock, N. (2015). Common method bias in PLS-SEM: A full collinearity assessment approach. *International Journal of e-Collaboration (IJEC)*, *11*(4), 1–10.

Lam, T., & Hsu, C. H. (2006). Predicting behavioral intention of choosing a travel destination. *Tourism Management*, *27*(4), 589–599.

Li, S. R., & Ito, N. (2021). "Nothing can stop me!" Perceived risk and travel intention amid the COVID-19 pandemic: A comparative study of Wuhan and Sapporo. In W. Wolfgang, C. Koo, & J. L. Stienmetz (Eds.), *Information and Communication Technologies in Tourism 2021* (pp. 490–503). Cham: Springer.

Liang, L. J., Choi, H. C., Joppe, M., & Lee, W. (2019). Examining medical tourists' intention to visit a tourist destination: Application of an extended MEDTOUR scale in a cosmetic tourism context. *International Journal of Tourism Research*, *21*, 1–13. doi:10.1002/jtr.2303

Liu, D., Li, R., & Johnson, S. (2016). Mainland Chinese outbound tourism to the United States: Recent progress. In X. R. Li (Ed.), *Chinese outbound tourism 2.0* (pp. 169–182). New York: Apple Academic Press. doi:10.1201/b19856

Mertens, G., Gerritsen, L., Duijndam, S., Salemink, E., & Engelhard, I. M. (2020). Fear of the coronavirus (COVID-19): Predictors in an online study conducted in March 2020. *Journal of Anxiety Disorders*, *74*, 102258. doi:10.1016/j.janxdis.2020.102258

Neuburger, L., & Egger, R. (2020). Travel risk perception and travel behaviour during the COVID-19 pandemic 2020: A case study of the DACH region. *Current Issues in Tourism*, *25*, 1–14. doi:10.1080/13683500.2020.1803807

Oglethorpe, J. E., & Monroe, K. B. (1987). Risk perception and risk acceptability in consumer behavior: Conceptual issues and an agenda for future research. In R. W. Belk et al. (Eds.), *AMA Winter Marketers Educators' Conference* (pp. 255–260). Chicago, IL: America Marketing Association.

Perić, G., Dramićanin, S., & Conić, M. (2021). The impact of Serbian tourists' risk perception on their travel intentions during the COVID-19 pandemic. *European Journal of Tourism Research*, *27*, 2705–2705. https://ejtr.vumk.eu/index.php/about/article/view/2125/466

Podsakoff, P. M., MacKenzie, S. B., Lee, J.-Y., & Podsakoff, N. P. (2003). Common method biases in behavioral research: A critical review of the literature and recommended remedies. *Journal of Applied Psychology*, *88*(5), 879–903. doi:10.1037/0021-9010.88.5.879

Preacher, K. J., & Hayes, A. F. (2008). Asymptotic and resampling strategies for assessing and comparing indirect effects in multiple mediator models. *Behavior Research Methods*, *40*(3), 879–891. doi:10.3758/brm.40.3.879

Quintal, V. A., Lee, J. A., & Soutar, G. N. (2010). Risk, uncertainty and the theory of planned behavior: A tourism example. *Tourism Management*, *31*(6), 797–805. doi:10.1016/j.tourman.2009.08.006

Rahimizhian, S., & Irani, F. (2020). Contactless hospitality in a post-COVID-19 world. *International Hospitality Review*, *35*, 293–304.

Rastegar, R., Seyfi, S., & Rasoolimanesh, S. M. (2021). How COVID-19 case fatality rates have shaped perceptions and travel intention?. *Journal of Hospitality and Tourism Management, 47*, 353–364.

Rather, R. A. (2021). Demystifying the effects of perceived risk and fear on customer engagement, co-creation and revisit intention during COVID-19: A protection motivation theory approach. *Journal of Destination Marketing & Management, 20*, 100564. doi:10.1016/j.jdmm.2021.100564

Reisinger, Y., & Mavondo, F. (2005). Travel anxiety and intentions to travel internationally: Implications of travel risk perception. *Journal of Travel Research, 43*(3), 212–225. doi:10.1177/0047287504272017

Rich, M. (2020). As coronavirus spreads, so does anti-Chinese sentiment. *The New York Times*, p. 30.

Ringle, C. M., Wende, S., & Becker, J. M. (2015). *SmartPLS 3*. Boenningstedt: SmartPLS GmbH.

Rittichainuwat, B. N., & Chakraborty, G. (2009). Perceived travel risks regarding terrorism and disease: The case of Thailand. *Tourism Management, 30*(3), 410–418. doi:10.1016/j.tourman.2008.08.001

Sánchez-Cañizares, S. M., Cabeza-Ramírez, L. J., Muñoz-Fernández, G., & Fuentes-García, F. J. (2020). Impact of the perceived risk from Covid-19 on intention to travel. *Current Issues in Tourism, 24*, 1–15. doi:10.1080/13683500.2020.1829571

Seow, A. N., Choong, Y. O., Moorthy, K., & Chan, L. M. (2017). Intention to visit Malaysia for medical tourism using the antecedents of theory of planned behaviour: A predictive model. *International Journal of Tourism Research, 19*(3), 383–393. doi:10.1002/jtr.2120

Shen, L., Cheng, C.-K., Lin, Y.-J., & Liu, H.-L. (2016). Mainland Chinese outbound tourism to Taiwan: Recent progress. In X. R. Li (Ed.), *Chinese outbound tourism 2.0* (pp. 93–116). New York: Apple Academic Press. doi:10.1201/b19856

Sinophobia: How a virus reveals the many ways China is feared. *BBC News*. (2020). Retrieved September 9, 2020, from https://www.bbc.com/news/world-asia-51456056.

Sparks, B., & Pan, G. W. (2009). Chinese outbound tourists: Understanding their attitudes, constraints and use of information sources. *Tourism Management, 30*(4), 483–494. doi:10.1016/j.tourman.2008.10.014

Steinmetz, H., Knappstein, M., Ajzen, I., Schmidt, P., & Kabst, R. (2016). How effective are behavior change interventions based on the theory of planned behavior?. *Zeitschrift für Psychologie, 224*(3), 216–233. doi:10.1027/2151–2604/a000255

Tavitiyaman, P., & Qu, H. (2013). Destination image and behavior intention of travelers to thailand: The moderating effect of perceived risk. *Journal of Travel & Tourism Marketing, 30*(3), 169–185. doi:10.1080/10548408.2013.774911

United Nations World Tourism Organization. (2020). *COVID-19 and tourism: Tourism in pre-pandemic times*. https://www.unwto.org/covid-19-and-tourism-2020

Vahedi, S., & Irani, F. (2021). The role of perceived tourists' susceptibility on perceived tourists' COVID-19 infectability among tourist arrival in North Cyprus. In Proceedings of MTCON'21, p. 231.

Wallace, L. E., Patton, K. M., Luttrell, A., Sawicki, V., Fabrigar, L. R., Teeny, J., et al. (2019). Perceived knowledge moderates the relation between subjective ambivalence and the "impact" of attitudes: An attitude strength perspective. *Personality and Social Psychology Bulletin, 46*, 014616721987349. doi:10.1177/0146167219873492

Wang, H. (2012). Value as a medical tourism driver. *Managing Service Quality: An International Journal, 22*(5), 465–491. doi:10.1108/09604521211281387

Part IV

Revisiting Chinese outbound tourism

17 Chinese outbound tourism in the post-SARS-CoV-2 era

Wolfgang Georg Arlt

Introduction

The second decade of the 21st century starts to appear in retrospect already as the final part of the "good old times". The Dow Jones index tripled between New Year's Eve 2009 and New Year's Eve 2019, the global GDP almost doubled in the same period, resulting also in a constant growth of GDP per person for all years of the decade. Even though the income gap between rich and poor continued to get wider, still the population growth slowed further down, and the percentage of persons living in poverty for the first time fell below 10%, while the literacy rate and the percentage of persons with access to sanitation and to telephone and internet connectivity all grew globally. The number of international trips grew from less than 1 billion in 2010 to almost 1.5 billion in 2019.

The negative effects of Climate change, the rise of populism and nationalism and the slowing of international integration all started to cast longer threatening shadows, but thanks to progress with alternative energy sources, carbon taxes and with innovations in IT and AI most experts forecasted for 2020 and the new decade more growth and prosperity (Kose & Ohnsorge, 2020). China prospered in the decade as well, with the GDP more than doubling and the number of persons living under the poverty line decreasing from 11% of the population to less than 1%. Environmental problems however increased dramatically with regard to air pollution and the depletion of the water table. In tourism, the number of domestic trips tripled from two billion to six trips, while China developed as the biggest, fastest-growing and highest spending international tourism source market, with the number of international trips tripling from 57 million in 2010 to 170 million in 2019. In that year, one out of nine international travellers carried a passport issued by the People's Republic of China.

The very success of international tourism in the 21st century however made a mockery of the idea of hospitality and ran a juggernaut over local nature, local culture, diversity and serendipity. In conferences and meetings, "Overtourism" took centre stage in the discussion about global tourism, as guests, hosts, service providers and environmentalists all saw their

DOI: 10.4324/9781003121329-21

satisfaction levels fall and tensions increase. However, most saw the problems as no more than a management problem of tourism "congestion" in specific urban areas (United Nations World Tourism Organization [UNWTO] et al., 2018), with a few exceptions: "Significant structural, cultural, environmental, economic, political changes are crushing the dogmas and paradigms that have persisted since Thomas Cook created the world's first travel agency in 1845" (Ascolese & Llantada, 2019, p. 28).

Chinese outbound tourists were often named as culprits for overtourism, with a lamentable lack of interest in the destination and a lack of sophistication on the one hand, but as "walking wallets" providing needed increases in arrivals and spending numbers on the other hand. Chinese travellers often felt that abroad, their needs and demands were not suffciently taken care of, whereas many staff members of hotels and tourism service providers as well as many locals ranked Chinese tourists among the least desirable.

According to UNWTO Secretary-General Zurab Pololikashvili, 2020 was anticipated with high expectations (Chan, 2019) for another good year for the "most resilient economic sector" (UNWTO, 2020). Then, just before the New Year 2020 was ushered in, a new form of a virus, Severe Acute Respiratory Syndrome Coronavirus 2 (SARS-CoV-2), which leads to the disease COVID-19, appeared. Starting as a Chinese problem in the weeks just before the start of the Chinese Year of the Yang Metal Rat, it quickly became a global problem, shutting down public life, production, transportation and also tourism for many months around the world in a way never witnessed before, including the most important international tourism source market China.

This chapter will discuss how the pandemic changed tourism and tourists in China and what can be expected from the new wave of Chinese outbound tourists once the borders open again sometime in 2022. Obviously with almost no Chinese travelling outside their own country when this text was written, the sources of the discussion will have to be mostly the travel behaviour in 2020 and 2021 in Chinese domestic travel as well as the results of surveys done during that time to measure the sentiment of Chinese potential international travellers. This text was finished by November 2021 and is reflecting the situation at that moment in the development. Quantitative data, if not mentioned otherwise, are based on the COTRI ANALYTICS database.

Tradition lost – Travelling for education, not for fun or adventure

Five hundred years ago, China under the Ming Dynasty and India under the Moghul dynasty were the two leading economies in the world, with a GDP at least five times as big as that of France under the Bourbons as a distant third. Both countries were the source of most inventions and progress in technology, medicine, arts and other fields, which found their way to Europe through Arabian and Central Asian intermediaries (Hansen, 2015).

Travelling for leisure, especially long distance, was mostly restricted to Asia until the advent of the Grand Tour in the 18th century in Europe. In China, if only considering its vast territory during many dynasties, travelling from one place to another was necessary for military, commercial, religious, representative, educational and administrative reasons. However, travelling was also seen as a positive – and possible – activity without a need to do so. Travelling was identified as a part of the personal education, of becoming a junzi (gentleman) and of realizing the majesty of nature compared to the smallness of human beings. Already the opening paragraph of the Analects, the most important book connected to the teaching of Confucius, starts with a sentence that positively mentions host-guest relations: "To study and at times put it into practice, is it not a pleasure? To have friends come from afar, is it not a joy?" (Lau, 1998).

Whereas the Bible in Genesis 1:28 says "Fill the earth and subdue it; and have dominion over the fish of the sea and over the birds of the air and over every living thing that moves upon the earth" (The Holy Bible - King James Version, 2021), Chinese traditional paintings depict small wanderers walking on narrow pathways among towering mountains, beside big rivers and under high heavens. Travelling is not to conquer or to entertain, but to learn and reflect.

Outbound travel to areas beyond the area of today's China, however, did not exist in China except for small colonies of Chinese migrants in neighbouring countries or exceptional trips like those of Buddhist monks to fetch Buddhist scriptures in India and famously by Admiral Zheng He's fleet to South Asia and East Africa to proclaim the overlordship of the newly established Ming Dynasty. After the fall of the Qing Dynasty in 1911, a brief period which witnessed a small group of rich urban Chinese copying Western leisure travel behaviour ended again soon with the occupation of the bigger part of China by Japan in the 1930s and early 1940s and the Civil War of 1945–1949.

Therefore, neither cultural pattern for international travel nor role models similar to Ibn Battuta, Marco Polo or Vasco da Gama existed. The first two decades of development of China's outbound tourism witnessed many complaints about the behaviour of the travellers being far removed from the ideals of Confucian and Daoist respect for nature and fellow humans and the idea of travelling as a learning and self-improvement process (Arlt, 2006).

From a Maoist point of view, leisure activities including domestic leisure travel were considered in the People's Republic of China as a wasteful, bourgeois practice. The wish to travel abroad was seen as practically equal to committing treason. With the beginning of the Reform and Opening policy and especially from the 1990s, inbound tourism was promoted as a fast way to earn hard currency. Domestic tourism became more and more acknowledged as an important part of the service industry, constituting a crucial element in the ideological switch from rural socialism to urban consumerism in China. However, outbound travel was permitted reluctantly and only step by step. A flexible response to the growing demand for international travel happened not the least under the impression of the Berlin Wall – and subsequently the Eastern Bloc communist governments – falling under the chants of the East Berliners not for *Freiheit* (Freedom) or democracy, but for *Reisefreiheit* (Freedom of travel). In any case, most of the long-distance trips in the 1990s and early 2000s were not leisure trips but government- or company-paid business trips by "delegations", with a smaller or larger touristic element included.

Only after the introduction of the Approved Destination System (ADS), self-paying travel for pleasure could be openly promoted. By the end of 2009, all major destinations had entered the ADS system or had special regulations like the SARS Hong Kong and Macau (Zhang, 2010).

At the same time, the Chinese government offcially declared tourism to be a "pillar industry" and started to consider outbound tourism as a "soft power" tool to demonstrate to the world the newly achieved might and power of China (Tse, 2013) and to promote or restrict visits of Chinese nationals to destinations in accordance with political goals (Arlt, 2015).

During the first decade of the new millennium, Chinese outbound travellers also developed into the biggest growth market in luxury brand shopping in all major destinations. For many luxury brands sales to Chinese customers, both at home and abroad became the most important segment of their overall sales. 2012 marked the year in which China not only reached the top position in terms of border crossings with 83 million trips but also with more than 100 billion USD in terms of spending, overtaking Germany and the USA as the biggest outbound tourism source market in the world.

The new Chinese government under Xi Jinping moved further away from trying to slow down and control the development towards emphasising its support for outbound leisure tourism by Chinese citizens. In 2014, he even urged Chinese tourists while visiting the Maldives "We should also educate our citizens to be civilized when travelling abroad. Don't litter water bottles, don't destroy their coral reef. Eat less instant noodles and more local seafood" (Gu, 2014).

In the same year, the USA announced that multiple entry visa for Chinese business and leisure travellers would increase in validity from one to ten years. This agreement was followed by many smaller countries, which even adopted policies of visa waiver or visa on arrival. By 2019, the Schengen region remained as almost the only major destination without such simplified policies for Chinese leisure travellers. With their increasing visibility and number, the image of Chinese tourists became more and more that of bad behaviour and ignorance both inside and outside of the country. Entering "Chinese tourists" into the search of Youtube.com resulted (and still results) an increasing number of links to videos showing outrageous behaviour of Chinese travellers.

These travellers represent the top decile of the Chinese society, as reflected in the fact that only about 12% of the population are holding a passport. Only members of the upper-middle class and the upper class are affluent enough to consider outbound leisure tourism paid by themselves. Most of them live in first (Beijing, Shanghai, Guangzhou, Shenzhen) and second-tier cities (Chengdu, Hangzhou, Chongqing, Wuhan, Xi'an, Suzhou, Tianjin, Nanjing, Changsha, Zhengzhou, Dongguan, Qingdao, Shenyang, Ningbo, Kunming). They are responsible for the biggest part of outbound departures from China, especially with regard to trips going beyond Hong Kong and Macau.

Already before the pandemic, about two-thirds of Chinese travellers did not fit anymore into the caricature of groups of large loud tourists spilling out of coach buses for short bouts of frantic picture taking while trampling the lawn, not speaking a word of a foreign language, buying mainstream luxury brands and insisting on Chinese food only (Ctrip & COTRI, 2018). Individual travellers and travellers in small groups with common interests had been making up a bigger part of the market with each year of growing experience and "live like a local" or at least "learn something about local culture" had moved up to the top of the list of demands, with language barriers lowered by more language skills as well as through technical solutions like translation apps. Whereas the sophistication improved, the quantitative growth both for the number of trips and the spending on outbound tourism dropped to single digit percentages, in line with the downward trend of China's GDP growth.

Chinese tourism in 2020 and 2021

The COVID-19 pandemic started with a number of pneumonia-like cases in Wuhan during the last days of 2019. Three weeks later, the "temporary" suspension of all group tours and package trips was announced, both within China and abroad (Bloomberg, 2020). In March 2020, the number of cases and deaths outside of China started to grow quickly, flights stopped, visa were cancelled and borders closed for Chinese travellers, to remain so for an unforeseen length of time. China also closed its borders to all foreigners at the end of March to stop the virus being returned into China from abroad.

In China, the pandemic was basically over in the middle of April 2020, with only four persons dying for COVID-19 between 18 April 2020 and the end of October 2021. It can be said that probably more Chinese died from accidents involving water buffalos than from the virus during that period of time (Newsflare, 2020). However, for a number of reasons, China followed a very strict zero-case policy and reacted with massive lock-downs and quarantine restrictions to outbreaks of only a few cases, even disrupting global supply chains by closing major harbours and factories for extended periods of time.

Domestically, China became soon the most active tourism region in the world again. On 4 April 2020, the first outings took place to national parks like the Huangshan Mountain during the May holidays, travelling between provinces became possible again without quarantine obligations following. Attractions like the Great Wall, the Forbidden City and many museums reopened, albeit with a limit to the number of visitors. In July 2020, the Ministry of Culture and Tourism issued a directive calling for the resumption of domestic tourism trade in all regions except those with a high coronavirus risk. Travel agencies and OTAs were allowed to start again to sell interprovincial group package tours. Travel during the October Golden Week 2020 was not discouraged, resulting in almost 700 million domestic trips.

Virus-free Macau SAR was opened for Mainland Chinese visitors in several steps in the second half of 2020 for visits without the need for quarantine after return. Between August 2020 and September 2021, about 7.5 million Mainland Chinese visited the SAR, making it not only the single outbound destination open for Chinese visitors but also one of the most-visited destinations in the world. Altogether, 2.8 billion domestic trips were been made by Chinese in 2020, 52% less than in the previous year. However, the total spending for cultural and tourism activities grew by 2% compared to 2019. Among the trends identified early on in the style of travel pattern were self-driving holidays, short distance trips and duty-free shopping (Global Times, 2021).

For the Chinese New Year Golden Week in February 2021, new cases of COVID-19 led to the governmental discouragement of travel during the holidays. Government staff were forbidden to travel, private employers paid bonuses to those staying and many home countries of migrant workers declared a need for entering a quarantine after arrival to discourage them from visits. Migrant workers living in bigger cities were encouraged to stay in the city rather than visit their families in rural areas, with free phone time and other incentives handed out to those staying in town. Even with these restrictions, the number of trips during the New Year period reached about half of the level of trips of the 2019 New Year period.

The Chinese citizens' strong wish to travel was further demonstrated during the three days' holiday in early April 2021 on behalf of the Qingming (Tomb sweeping) festival. The number of trips reached 93% of the level of 2019; however, due to steep discounts for accommodation and entrance tickets, the level of spending reached only about 60% of the 2019 level. In May 2021, the number of trips even exceeded the number in 2019, while spending was still below the pre-pandemic level. For the Golden Week in October 2021, the travel restrictions and the official discouragement of the government to go on leisure journeys had increased again following some minor outbreaks, but the number of domestic trips still stood it 70% of the level of 2019, the spending however stayed at 60% of the 2019 numbers, a result of the growing importance of self-drive trips to destinations within the region (Arlt, 2021a).

China's post-pandemic tourism

When the time of domestic-only tourism will end, remains in the autumn of 2021 the one-million-dollar question (Wang, 2021). In June 2021, the *Wall Street Journal* scared the global tourism community by forecasting that the borders would remain closed until after the re-election of Xi Jinping as Secretary General of the Chinese Communist Party in November 2022 (Zhai & Hua, 2021). An article by McKinsey even put the restart of China's outbound tourism to the first half of 2023 (Chen et al., 2021). The fact that China's government decided that it will not welcome foreign spectators for the Beijing 2022 Winter Olympics and Paralympics, which are scheduled for February and March, was seen as an indicator that outbound travel will not start already for the Spring Festival 2022.

However, an article in the *China Daily*, which appeared on 12 October 2021, had for the first time a government official saying that the borders could be opened again at the beginning of 2022:

> China may open its borders after it vaccinates over 85 percent of its population by early 2022, said Gao Fu, head of the Chinese Center for Disease Control and Prevention, in a recent interview. The current epidemic control measures against COVID-19 in China have won a lot of time to produce enough vaccines and inoculate people, he told a news outlet. If the vaccination rate can reach more than 85 percent in early 2022, there will be few infections and none of those infected would be in serious condition or die. Besides, the virulence of the coronavirus is also decreasing, according to Gao. 'By that time, why shouldn't we open up?' he said.
>
> (Chen, 2021)

The spin to the story seemed to change, as the virulence of the SARS-CoV-2 is decreasing only in terms of serious illnesses in a population which is mostly vaccinated, whereas the virulence for an individual unvaccinated person remains the same. It can be speculated if this reference to statistical trickery is a sign of the Chinese government admitting that the zero-case policy does not work and that the affluent citizens should be given the freedom of international travel again as a safety valve for any displeasure after the recent power shortages, internet restrictions and looming wealth taxes? After all, in October 2021, only Turkmenistan and North Korea were still insisting on maintaining the zero-case policy (Arlt, 2021b).

When discussing the future travel behaviour of Chinese outbound tourists, it becomes clear that as a result of the pandemic, the Chinese travel demand and expectations have evolved, and trends existing before have been accelerated but also new trends have emerged, that will eventually influence the demand and behaviour of Chinese outbound tourists after the travel restrictions will be eased.

Among the prevailing demands, several will continue to be of importance. Seeing the world with one's own eyes, sampling new experiences is still a major attraction given the limitations of access to free and unbiased information in China and the recent start of the Chinese outbound travel.

Gaining social prestige, confirming status and self-image continues to be important within the Chinese society, with international travel as a major tool to achieve such an objective similar to the ownership of a big apartment, an expensive car, watch and clothes. With many Chinese following a serious leisure subject (Stebbins, 2007) and/or having several hobby interests, outbound trips can be used to realize such demand for photography, sports, gastronomy, music, traditional transportation forms and many others.

Travelling in the footsteps of peers and idols has been an important aspect of the decision-making process in the selection of destinations. However, the so-called Key Opinion leaders (KOLs), celebrities including singers, actors and industry leaders who advertise products and themselves in partnership with luxury brands and destinations, have fallen out of favour with many Chinese consumers who have learned to identify commercially driven recommendations as based on the stars' lucrative marketing deals rather than on their real preferences. Instead, Key Opinion Customers (KOCs) are gaining influence. These influential consumers help to "seed" products or destinations into communities and groups they are part of WeChat or other platforms. A KOC is primarily a customer with a very efficient organic reach within family members, acquaintances and colleagues in his or – more often – her closed private network. KOCs can strongly act for their peers as "influencers" through their objectivity in opposition to the paid KOLs. As China is a culture based on networks, KOCs, in most cases, have a range of 1,000 to 5,000 or more peers.

The pandemic years have accelerated the shift from KOL to KOC marketing. Online reviews are highly important for the decision-making process of more than two-thirds of Chinese consumers according to several recent surveys. Reviews done by KOCs are seen by most Chinese as providing authentic and reliable content and therefore as more trustworthy than those by paid KOLs. Some experts even see a shift to another new kind of influencer in the form of Cultural Opinion Leaders, engaging with the 250 million Chinese Gen Z consumers on platforms like Xiaohongshu, Bilibili and Poizon (Achim, 2021).

Except for pensioners, before and after the pandemic, most Chinese outbound travellers had and will have more money than time, given the lack of established paid holiday systems in China. With a long bucket list of must-see places and activities, they tend to prefer to do many different things in a rather short time.

As a result of the forced concentration on domestic tourism, international destinations need to work on their competitiveness, as the interest in domestic destinations has risen considerably, making domestic Chinese destinations an additional new competitor. Travelling inside China is no longer seen as a sign of lack of money and sophistication, but – encouraged by the Chinese government – as "patriotic" tourism, supporting the national tourism industry (O'Farrell, 2021). Travelling to support the appreciation of the native home as one element of a nationalistic approach, together with buying Chinese instead of international brands in line with the "Guochun" movement and watching Chinese instead of Hollywood movies, brings the perception of travel full circle back to the days of Confucius.

In the short term, Chinese tourists expect destinations to be safe, prepared and able to handle any crisis. This includes, for example, arrangement of transportation in emergency situations and flexible cancellation policies. Favoured destinations are those that can ensure the safety of their visitors, have less crowded places and plenty of nature and attractions supporting health and family time. This will also be valid for international travel after its restart. As live streaming is becoming more and more popular, and the Chinese were able to have virtual tours of destinations all over the world, there is an increased demand for nature-focused travels and outdoor activities. In the same line is the interest in visiting rather smaller cities and new destinations than big metropoles.

The safety aspect however does not only connect to medical topics in outbound tourism. With China standing accused of being the source of the pandemic or of even wilfully spreading it, more importance than ever is given to the question if in a given destination Chinese as welcome and treated in a friendly way. The incidents of "Asian Hate" attacks on Chinese-looking persons especially in the USA have been given a lot of attention in Chinese media, claiming that Asian Americans "are more stressed by racism than by the pandemic" (Zhu, 2021).

The pandemic has accelerated a further increase in interest in independent international travel rather than package tours. Furthermore, exclusivity (luxury hotels, gourmet private groups), adventure (diving, paragliding) and nature experience (outdoor hiking, hot air ballooning) will find interest, albeit with different Chinese market segments (Wang & Parulis-Cook, 2021).

Differentiation is important to understand the changes in the Chinese tourism market, as obviously "the" Chinese outbound tourist does not exist and that different market segments have reacted differently to the situation in 2020 and 2021. A 55-year-old granny from a rural part of China, which might have been invited to an outbound trip by her children, travelling abroad with a package tour for the first time, a 33-year-old real estate agent from Shanghai with a degree from an Australian university on his honeymoon trip and a 22-year-old female students who grew up in an affluent family visiting together with a few other female fellow students places which were featured in the "Game of Thrones" TV series (Wang, 2019), are all obviously very different people with very different needs and expectations. They will all share some characteristics like the lack of time and the insistence of respect being paid to all things Chinese, but their level of information, consumption patterns and preferred level of involvement with local culture and inhabitants were distinctively different before and will be even more so after the pandemic is over.

In terms of residence, as most affluent Chinese live in first (Beijing, Shanghai, Guangzhou, and Shenzhen) and second-tier cities, they also represent the majority of travellers especially for long-distance trips. The fastest growth rates in the second half of the last decade, however, resulted in third and lower-tier city dwellers entering the market, albeit mostly to Asian destinations.

The travel behaviour and preferences of Chinese tourists from the different level tier cities are different ranging from accommodations to activities. Better educated, richer and more internationally experienced people are most likely to live in first-tier cities, to speak some English and to have more interest than before in meeting local interesting persons with whom they share interests and in immersion into a local community, if only for a short period of time. The more important second-tier cities like Chengdu, Chongqing, Hangzhou and several others, however, also managed to attract a growing number of high achievers in recent years. They are much more likely to travel in smaller groups of friends, family or colleagues or as FITs, increasingly as self-drive trips with rented limousines or RVs.

The majority of the Chinese preferring package tours are from second or lower-tier cities. Catching up swiftly, they have overtaken the number of first-tier citizens in total outbound travel, but not in long-haul trips yet. Those who have yet to gain more experience with travel abroad are usually travelling in typical Chinese package tour groups which include a tour guide and a fixed itinerary, bringing together strangers in one travel group. However, even such tours that typically rush to tick off the major sights in as many destinations as possible, persist on Chinese food and spend a lot of time – encouraged by the tour guide with shopping will further change after the pandemic as the bragging power gained from such trips is on the vane even within less sophisticated peer groups.

Many package tour participants are only interested in getting a taste of the local food and go back to Chinese food to fill their stomachs, even if the quality and choice between different Chinese cuisines is in most cases sub-standard, especially outside of Asia. Many fear that they might end up with digestion problems. It should be noted, however, for Starbucks, McDonalds, KFC, Taco Bell, etc. China is the most important market outside of the USA, so that the offers of the system gastronomy are not "foreign" anymore for Chinese.

For most of the more experienced Chinese travellers, however, eating authentic local cuisine is one of the main requests. Becoming a knowledgeable "foody" or a wine or craft beer expert will be considered an important badge of honour to acquire. Sharing such experiences via social media channels will increase their standing among their peers. However, what is considered as "authentic" is often more influenced by Chinese bloggers than based on real local traditions. Storification, rituals and costumes further add to the attractiveness of the local food offered (Arlt, 2018).

Most of the more experienced travellers are increasingly eager to visit and explore new emerging destinations and to have authentic experiences with the local cuisine, nature and people. During the pandemic, they had a lot of time and opportunity to amass information also about less famous destinations and attractions, supported by the marketing activities of museums providing virtual 3D tours, videos and many other forms of information provided by NTOs and DMOs eager not to lose touch with the potential customers. This results in the continuous growth of the Chinese outbound segment of FITs in smaller groups, such as travelling as a couple, with friends or family, or specific "semi-self-organized" tours that are customized to the demand of the Chinese tourist and organized by tour operators.

The concept of consumerism for Chinese tourists was already changing before the pandemic and will certainly change even more in post-pandemic times. Not only are most foreign products available in China, buying especially bigger luxury brand goods is not increasing the cultural capital of the shopper in the eyes of her (or his) peers anymore to the same degree as before the pandemic. The focus of spending will further move towards buying high-quality local resources during travels, which are unique or at least not available in China and, if possible, connected to a specific story and shopping experience. A bigger percentage of the travel budget is furthermore spent on fine-dining and on specific experiences, often connected with pursuing a specific hobby as a form of serious leisure.

The overall level of spending per trip of per person per day during the trip varies widely, the average spending per person per day, however, will further decrease for several reasons:

The proportion of shopping for luxury brand goods during the trip decreases as more duty-free shopping offers especially on Hainan and on new online shops in China have become available. The number of Chinese buying goods based on a shopping list given to them by friends and neighbours is decreasing as these persons are more likely to travel themselves, so that the same amount of money spent on luxury brands is now split between the travellers.

The volume of spending of "daigous", who buy goods in bulk for resale in China, will further decrease as more Chinese are travelling themselves, price differences are not as big as before between China and international markets, and because the Chinese government is controlling and taxing the daigou business more than before. The anti-corruption campaigns of the Xi government have reduced the amounts of illicit funds which can be legalized through the relatively simple money laundering tool of shopping abroad (Arlt, 2021a).

In terms of age, after the "First Wave" of package tour travellers and the "Second Wave" (Arlt, 2013) of FITs, some experts have identified a "Third Wave" (Bao et al., 2019), referring to the age group above 55 years, which is growing faster than the majority of young and middle-aged travellers. The first cohorts of affluent elderly entering the market as well as the increase in families travelling abroad as three-generation groups are mainly responsible for this development.

The first Chinese pensioners with enough wealth started to travel before the pandemic, strengthening the "Best Agers" or "Silver Travellers" segment with most of them being between 55 and 65 years old. According to the census results published in May 2021, the number of Chinese aged 60 and above rose to 264 million, representing 19% of the total population. Those with sufficient money and time to spend tended to go on longer trips that could even last several weeks and can be expected to do so even more in the coming years. This group is especially interested in culture and history and on the lookout for off-the-beaten-track destinations and able to travel outside of the main season.

As a result of 30 years of one-child policy, the 2020s and 2030s will witness a strong increase in the number of older Chinese citizens, whereas the number of children and teenagers will continue to decrease due to the smaller number of young women. Even after the end of the one-child policy, many young Chinese do not want to marry, or if so, still do not plan to have more than one child (Xinhua, 2021).

In comparison, the majority of middle-aged Chinese tourists that have the budget to travel are not yet pensioners and cannot expect to inherit bigger amounts of money. This market segment has less time to travel and hence pays a lot of importance to making the best out of the limited time available. They further demand their host destination to treat them specifically with respect, as they are aware of the importance that Chinese tourists can have for a destination. As a sign of respect, they require information in Chinese even if they do understand English and want to be set apart from other tourists through a clear and distinct better treatment. Additionally, as most outbound travellers are working hard at their jobs throughout the year, they tend to desire intense and fast experiences without waiting time between activities. After the pandemic, a bigger percentage of this market segment will however also change the travel habit towards a more meaningful travel behaviour, slowing down and showing more interest in the host community.

Conclusion

To summarize, in the post-pandemic part of the 2020s, the Chinese outbound tourism market will continue to develop and diversify. Already, demand for new form of travelling and for new markets and activities are becoming evident in the Chinese market. For service providers and destinations, the challenge is to provide customized products fitting with the growing level of demand and the ongoing segmentation of the market. Niches in China have the positive characteristic to even if less than 1% of the population are interested in a specific theme or activity, this still translates into hundreds of thousands or millions of potential customers in absolute numbers, who are more likely to be part of the upper end of the society, affluent enough to turn their interest into actual travel.

The pent-up demand of the pandemic will translate into rapidly increasing numbers of outbound travellers once the hurdles have been removed, but it will be different from the Chinese outbound tourism before the pandemic (Arlt, 2021c). Nature travel, outdoor activities in summer and winter, adventure and sport trips, glamping and RV trips, fine-dining, honeymoon and wedding photo trips as well as all forms of acquiring skills in arts and gastronomy will become increasingly attractive especially to the more affluent parts of the Chinese market. Concentrating on quality instead of quantity, on duration of stay, satisfaction and yield rather than arrival numbers, will create a more sustainable and meaningful Chinese outbound tourism and will induce recommendation marketing by the satisfied visitors and guests.

References

Achim, A.-L. (2021, October 25). The birth of China's cultural opinion leaders. *Jing Daily*. Retrieved October 22, 2021, from https://jingdaily.com/china-cultural-opinion-leaders-influencers/

Arlt, W. G. (2006). *China's outbound tourism*. London: Routledge.

Arlt, W. G. (2013). The second wave of Chinese outbound tourism. *Tourism Planning & Development*, *10*(2), 126–133. doi:10.1080/21568316.2013.800350

Arlt, W. G. (2015). China's outbound tourism: History, current development and outlook. In R. Lee (Ed.), *China outbound tourism 2.0* (pp. 3–20). Oakville: Apple Academic Press.

Arlt, W. G. (2018). Chinese outbound tourists: Food and beverages. In C. Pforr & I. Phau (Eds.), *Food, wine and China: A tourism perspective* (pp. 47–59). London: Routledge.

Arlt, W. G. (2021a). New trends in Chinese outbound tourism: Consequences for the international hospitality industry. In M. A. Gardini, M. C. Ottenbacher, & M. Schuckert (Eds.), *The Routledge companion to international hospitality management* (pp. 211–216). New York: Routledge.

Arlt, W. G. (2021b, October 19). *COTRI WEEKLY Editorial October 19th – Spin doctors taking over* [Video]. YouTube. https://www.youtube.com/watch?v=xyQGvDccjYQ

Arlt, W. G. (2021c, May 27). Indoor glamping and RVing – Welcome to post-pandemic outbound tourism. *4Hoteliers: Hotel, Travel & Hospitality News*. Retrieved October 22, 2021, from https://www.4hoteliers.com/features/article/14453

Ascolese, G., & Llantada, J. (2019). *The next great tourism revolution: A report on travel and tourism trends*. Valencia: WAM. https://www.wearemarketing.com/uploads/media/default/0001/21/be54a5f27191d02797ff03aaacd4b799f25beb11.pdf

Bao, J., Jin, X., & Weaver, D. (2019). Profiling the elite middle-age Chinese outbound travellers: A 3rd wave? *Current Issues in Tourism*, *22*(5), 561–574. doi:10.1080/13683500.2018.1449817

Bloomberg. (2020). China orders travel agencies to suspend tours to contain virus outbreak. *Bloomberg*. Retrieved October 22, 2021, from https://www.bloomberg.com/news/articles/2020-01-24/china- orders-travel-agencies-to-suspend-all-tour-sales

Chan, A. (2019, July 26). China outbound tourism trends prediction 2022. *Travel Daily News*. Retrieved October 22, 2021, from https://www.traveldailynews.asia/china-outbound-tourism-trends-prediction-2022

Chen, G., Phillips, M., Saxon, S., & Yu, J. (2021, August 30). China's uneven travel recovery: Long road to international travel furthers domestic opportunities. *McKinsey & Company*. Retrieved Oct. 22, 2021, from https://www.mckinsey.com/industries/travel-logistics-and-infrastructure/our-insights/chinas-uneven-travel-recovery-long-road-to-international-travel-furthers-domestic-opportunities

Chen, M. (2021, October 12). Borders might open if vaccination target reached by early 2022. *Chinadaily.com.cn*. Retrieved October 22, 2021, from https://www.chinadaily.com.cn/a/202110/12/WS61659479a310cdd39bc6e79f.html

Ctrip & COTRI (Eds.). (2018). *Customised travel report: 2017 trends in Chinese customised travel to Europe*. Shanghai: Ctrip.

Global Times. (2021, July 5). Chinese domestic tourism trips down 50% in 2020: Official data. *Global Times*. Retrieved October 22, 2021, from https://www.globaltimes.cn/page/202107/1227854.shtml

Gu, L. (2014, September 17). 'Eat more seafood and less instant noodles,' Xi suggests Chinese tourists. *Ecns.cn*. Retrieved October 22, 2021, from http://www.ecns.cn/cns-wire/2014/09-17/134894.shtml

Hansen, V. (2015). *The open empire: A history of China to 1800*. New York: W. W. Norton & Company.

Kose, M. A., & Ohnsorge, F. (2020, February 20). Emerging and developing economies: Ten years after the global recession. *CAMA Working Paper No. 13/2020*. doi:10.2139/ssrn.3541418

Lau, D. C. (Ed.). (1998). *The analects: Sayings of Confucius*. London: Penguin Classics.

Newsflare. (2020, April 19). Runaway buffalo attacks woman holding a child at front of her house in southern China. *Newsflare*. Retrieved October 22, 2021, from https://www.newsflare.com/video/361570/runaway-buffalo-attacks-woman-holding-a-child-at-front-of-her-house-in-southern-china

O'Farrell, S. (2021, July 12). Chinese tourists stay home. *fDi Intelligence*. Retrieved October 22, 2021, from https://www.fdiintelligence.com/article/79919

Stebbins, R. (2007). *Serious leisure: A perspective for our time*. New Brunswick, NJ: Transaction Publishers.

The Holy Bible - King James Version (K. J. Bible, Trans.; Kindle ed.). (2021). London: HarperCollins.

Tse, T. S. M. (2013). Chinese outbound tourism as a form of diplomacy. *Tourism Planning & Development*, *10*(2), 149–158. doi:10.1080/21568316.2013.783738

United Nations World Tourism Organization. (2020). *International tourism growth continues to outpace the global economy*. https://www.unwto.org/international-tourism-growth-continues-to-outpace-the-economy

United Nations World Tourism Organization, Centre of Expertise Leisure, Tourism & Hospitality, NHTV Breda University of Applied Sciences, & European Tourism Futures Institute of NHL Stenden University of Applied Sciences. (2018). *'Overtourism'? – Understanding and managing urban tourism growth beyond perceptions: Executive summary*. Madrid: UNWTO. https://www.e-unwto.org/doi/epdf/10.18111/9789284420070

Wang, M. (2019, May 22). Tourists take to Game of Thrones trail. *Chinadaily.com.cn*. Retrieved October 22, 2021, from https://global.chinadaily.com.cn/a/201905/22/WS5ce5081ea3104842260bd296.html

Wang, M. (2021, October 19). European tourism pins hopes on vaccinated visitors. *Chinadaily.com.cn*. Retrieved October 22, 2021, from https://www.chinadaily.com.cn/a/202110/19/WS616e2fc1a310cdd39bc6fc9b.html

Wang, M., & Parulis-Cook, S. (2021, September 23). China traveler sentiment report: September 2021. *Dragon Trail*. Retrieved October 22, 2021, from https://dragontrail.com/resources/blog/china-traveler-sentiment-survey-9-2021

Xinhua. (2021, May 12). Latest census reflects China's changing demographic landscape. *Xinhuanet*. Retrieved October 22, 2021, from http://www.xinhuanet.com/english/2021-05/12/c_139939402.htm

Zhai, K., & Hua, S. (2021, June 22). China to keep COVID-19 border restrictions for another year: Beijing worries about variants as it stages the Olympics and a power transition. *The Wall Street Journal*. Retrieved October 22, 2021, from https://www.wsj.com/articles/china-to-keep-covid-19-border-restrictions-for-another-year-11624361777

Zhang, G. (2010). China's tourism development: Situation, analysis and future perspectives. In W. G. Arlt (Ed.), *COTRI Yearbook 2010* (pp. 9–30). Heide: COTRI.

Zhu, L. (2021, June 1). Reports: Asian Americans more stressed by racism than pandemic. *Chinadaily.com.cn*. Retrieved October 22, 2021, from http://global.chinadaily.com.cn/a/202106/01/WS60b59a16a31024ad0bac2c41.html

Index